REVOLUTIONS

PAVING THE WAY FOR THE BIOECONOMY

REVOLUTIONS
PAVING THE WAY FOR THE BIOECONOMY
Randall E. Mayes

Published in The United States of America
by
Logos Press®, Washington DC
WWW.LOGOS-PRESS.COM
INFO@LOGOS-PRESS.COM

10 9 8 7 6 5 4 3 2 1

ISBN-13
Softcover: 978-1-934899-24-3

Library of Congress Cataloging-in-Publication Data

Mayes, Randall E. (Randall Elam)
  Revolutions : paving the way for the bioeconomy / Randall E. Mayes.
      p. cm.
  Includes bibliographical references and index.
  ISBN 978-1-934899-24-3 (softcover)
  1. Genomics. 2. Bioengineering. I. Title.
  QH447.M3528 2012
  572.8'6--dc23
                        2012017613

REVOLUTIONS
PAVING THE WAY FOR THE BIOECONOMY

Randall E. Mayes

LOGOS PRESS

Contents

1. What Genomics Revolution?   1

2. Mining Genomes   19

3. Out of the Woodwork   39

4. The Rise & Descent of Vitalism   55

5. How the Giraffe got its Neck   65

6. Four Waves   79

7. Crossing the Weismann Barrier   99

8. The Genomics Bubble   107

9. What can Economists Learn From Evolutionary Biology?   123

10. The Free-Rider Problem   137

11. The Boo Scenarios   147

Notes and Sources   161

Index   175

Acknowledgements

With the recent explosion of information in biology and genomics, the assistance of reference librarians made my research much easier. For gathering research materials, I express thanks to Duke University librarians and support staff Carlton Brown, Jane Day, Ben Fisher, Stephanie Ford, Teddy Gray, Joy Hanson, Sarah Hodkinson, Carson Holloway, Seth McCurdy, Dave Munden, Lee Sorenson, and Meg Trauner; and also to Nancy Kozlowski of the North Carolina Biotechnology Center who assisted me with navigating through repositories of cutting edge research in on-line databases.

For interpreting this information, I turned to correspondence and discussions with Phil Benfey, Rob Carlson, Drew Endy, Timothy Haystead, Chris Holman, Randy Jirtle, Daniel Kevles, Seymour Mauskopf, Alex Rosenberg, Pardis Sebati, Kevin Schulman, Michael Waitzkin, Spencer Wells, James Wyngaarden, and Lingchong You.

For advice on effectively communicating this information, I am grateful to Duke University rhetorician George Gopen.

Epigraph

Commenting on the relationship between bioethics and the Constitution: The tools for this job are not to be found in the lawyer — and hence not in the judge workbox.

Justice Antonin Scalia, 2004[1]

Acquired traits pass into the next generation (now known as the second law of Lamarckian inheritance).

Jean Baptiste Pierre Lamarck, 1809[2]

The transmission of acquired characters is impossible, because the germ plasm is derived from that which preceded it.

August Weismann, 1892[3]

In the excitement following the pursuit of the possibility that information resided in DNA sequences, alternatives were ignored.

Sahotra Sarkar, 1996[4]

Nature works as a tinkerer with available materials, not as an engineer does by design.

Francois Jacob, 1977[5]

Preface

Futurists have touted the 21st century as the century of biology. This is primarily due to the potential of genomics. Futurists have based that potential on anticipated revolutions. These revolutions present themselves in many fashions — industrial, scientific, cultural, and Schumpeterian. Revolutions: Paving the Way for the Bioeconomy is an in-depth look at these revolutions.

INDUSTRIAL REVOLUTIONS

Genomics researchers hope to launch personalized medicine and cure diseases by identifying drug targets and create novel therapies such as DNA vaccines by discovering gene variants that are risk factors called biomarkers. This new approach to medicine will ideally provide a boost to the biotechnology industry. So far, genomics has provided new tools for conducting biological research and more powerful tools for managing and interpreting data (bioinformatics).

SCIENTIFIC REVOLUTIONS

Although Darwin was unable to provide a mechanism for evolution via natural selection, science historians credit him with making evolution a believable concept supplanting vitalism. Eventually, collaborations leading to The Modern Synthesis were able to unify Mendelian genetics, natural selection, and mutations in to a quantifiable mechanism for understanding simple phenotypes.

Through genomics, many new discoveries in basic science have taken place. As a result, we are currently experiencing a Kuhnian revolution, a revolutionary change in how a majority of scientists view the world. Through epigenetics, evo-devo, and other discoveries in molecular biology, scientists have a new understanding for the concept of a gene. With the resurgence of the Weismann-Lamarck debate regarding mechanisms of developmental biology, niche construction has also emerged as a mode of inherited traits.

Using DNA, evolutionary anthropologists have demonstrated that in addition to genes and the environment, culture and technology also contribute to phenotypes. Rather than genetic determinism, the systems approach is now the norm in the experimental design of biological research.

CULTURAL REVOLUTIONS

Genomics is currently facing regulatory policy issues in the areas of risk assessment, intellectual property, and bioethics. Whether or not citizens reap the social goods and economic benefits from biotechnology will depend on the actions taken by activists, lobbyists, scientists, and the government. To receive the social goods and economic benefits from genomics, public acceptance is critical. It is important that the public understands and accepts that culture and technology have played an important role in what makes us human.

SCHUMPETERIAN REVOLUTIONS

The anticipated proof of concept in developing cures to complex diseases through genomics has yet to materialize. Using genomics to discover treatments and subsequent cures for diseases is more complex than originally thought. Instead, the field is currently experiencing a genomics bubble. Without genomics providing a proof of concept for medical cures, a paradigm shift for understanding diseases, consequently an economic (Schumpeterian) revolution in the pharmaceutical industry has yet to occur.

Synthetic genomics, more commonly referred to as synthetic biology, is poised to emerge as the next industrial revolution and has provided revived optimism. Using synthetic biology, bioengineers can reprogram DNA or create novel DNA and potentially make products such as vaccines and fuels solving major world problems.

CHAPTER 1

What Genomics Revolution?

In 1920, German botanist Hans Winkler coined "genome," ccombining gene and chromosome, referring to genes within a chromosome. In that time period, researchers were interested in hybrids that have distinctive genes. This was not a novel idea, however, as humans have crossbred plants and animals for thousands of years coinciding with their domestication. The numerous varieties of pet dogs available today are the result of are the crossbreeding of wolves with wild dogs resulting in traits for both companionship and protection.

For centuries, farmers have crossed donkeys with horses producing mules that have the beneficial traits from species, hardiness and intelligence. Through the 1900s technical improvements using the understanding Mendel's laws of genetics and mutations, the modern science of breeding began. This led to the eugenics movement not only for improving humans, but for the commercialization of plants and animals.

The U. S. Congress passed legislation in 1930 providing patent protection to plant breeders using asexual processes including grafting, budding, and roots and in 1970 for seeds used for horticultural and agricultural purposes. A Supreme Court ruling in the 1980s on genetically modified bacteria opened the door for animal patents. Today, scientists use artificial insemination and are investigating the cloning of race horses and using molecular biology techniques to create dairy and beef cattle that are more resistant to drought and produce healthier meat.

These scientific advancements have raised activist concerns on a number of levels. With each application of advances in science, the government is placed in the position of weighing the societal benefits with a new set of public policy issues related to bioethics, intellectual property, and risk assessment of public safety to humans and the environment.

In the early 1970s, medical researchers became interested in isolating mutated genes and exploring the potential of gene therapy for treating diseases. The process involves altering or removing the mutated gene and inserting a functional gene. The field experienced some success in clinical trials, but in some cases patients experienced problems with an immune response. Also, the treatment is temporary and viral vectors typically produce side effects in patients such as toxicity or mutations leading to tumors.

Diseases caused by mutations in a single gene including sickle-cell anemia, hemophilia, muscular dystrophy, Huntington's disease, and cystic fibrosis are candidates for gene therapy. However, most diseases are polygenic, or linked to more than one gene. Mary Claire King and Francis Collins were the pioneers in identifying genetically linked disease genes. In the early 1980s, these researchers had small sequencing projects where they cloned and stored fragments of DNA.

In 1982, the Department of Energy's (DOE) Los Alamos National Laboratory made sequencing data available in the National Institutes of Health's (NIH) public online database GenBank. But, genomics was still not a commonly used word; however, that would certainly change. In the mid-1980s, researchers began independently discussing and proposing mapping and sequencing the human genome to help the science community better understand disease mechanisms and basic science.

These discussions eventually led to the Human Genome Project, which became two separate efforts, consisting of public and private versions, to sequence the entire human genome. The dynamics of the two projects provided a melodramatic story with the characters sometimes overshadowing the plot, providing the makings of a Hollywood script. If a screenwriter prepared such a script, Act I would comprise the discussions and meetings by scientists to assess the feasibility of a large-scale government project; Act II would chronicle the public Human Genome Project; and like the Sputnik satellite, a private project provided a rivalry and a race soon began setting the stage for Act III.[1]

ACT I: DISCUSSIONS AND MEETINGS

In the 1970s Stanford University geneticist David Botstein realized the need for a map with DNA markers that would aid in locating disease genes. A chromosome map of a species shows the position of its known genes and markers relative to each other, rather than as specific physical points on each chromosome. Genetic distance is measured centiMorgans (cM) or approximately one million bases in honor of Thomas Hunt Morgan. Researchers set a goal to have markers every cM, totaling 3100 markers. In 1996, scientists surpassed that goal completing a map with 5,264 AC/TG unique sequence repeats called microsatellites as markers.[2]

In 1978, David Botstein found that using restriction enzymes produced different fragments in individuals, referred to as restriction length fragment polymorphisms (RFLPs). The genetic variation in the length of DNA fragments is useful as markers. These specific polymorphic sequences in the genome vary between individuals in a population sometimes as a result of a single base pair mutation. RFLPs occur at cleavage sites for their specific restriction enzymes and, the sequences revealed by RFLPs provide a starting point for the isolation of genes.

Tumor virus researchers including Renato Dulbecco were interested in genome sequencing to better understand cancer. Dulbecco believed sequencing would lead to the genes involved in cancer's progression. In an editorial for the journal *Science*, Dulbecco wrote, "We have two options discovering the genes involved with malignancy, the piecemeal approach or by sequencing the whole genome of a selected animal species."[3]

After James Wyngaarden became the director of the NIH in 1982, the former purine researcher and chairman of Duke University's Department of Medicine expressed interest in diagnostic and therapeutic advances possibly exploited from the human genome. However, some biologists were opposed to a project since only a small percentage of the genome was presumably genes. In contrast, Wyngaarden was curious what we could learn from non-coding DNA.[4]

Discussions of a large scale project led to a number of meetings and conferences to assess its feasibility. In 1985, molecular biologist Robert Sinsheimer organized a meeting in Santa Cruz, California where the idea was rejected, but not forgotten. In 1986, the DOE held a conference in Santa Fe, New Mexico, and James Watson held a meeting at Cold Spring Harbor Laboratory. In 1986, Charles DeLisi with the DOE explored the feasibility of DNA sequencing for detecting induced mutations in survivors of Hiroshima and Nagasaki and also for birth defects.

According to Leroy Hood, the participants of these meetings were evenly split on whether the Human Genome Project was a good idea. Those that were skeptical were not so much concerned about the science and the development of technologies for analyzing data, rather its practicality. Since government funded projects require Congress to authorize the budget, scientists were understandably concerned with the costs.

In the 1970s and 1980s, several researchers developed prototype automated DNA sequencing machines. Walter Gilbert of Harvard developed a machine using chemical methods. At the time of the 1985 meetings to discuss sequencing the human geneome, sequencing a single base cost $10 and DNA sequencers could read only 50 to 100 bases per day. Leroy Hood's lab at the California Institute of Technology developed machines using radioactivity which were capable of sequencing 200-300 base pairs per reaction. Caltech licensed the technology to Applied Biosystems Inc. which introduced them on the market in 1987. Today, researchers use the Sanger method which is easier to automate and named after British biochemist Frederick Sanger who invented the sequencing technology.

Perhaps the major concern was who could best manage a large scale project. A National Academy of Sciences study recommended mapping, sequencing, and further understanding the genome, but did not suggest which agency should lead. A number of scientists including David Baltimore were not convinced NIH was the agency that should lead the project. There were reservations with NIH involvement with big or managed science and with the adequacy of existing databases for handling all the generated data.[5] Some scientists and government officials suggested that mass sequencing is more appropriate for the DOE which had more centralized management.

In the 1980s the culture of NIH was small science with individual investigators. The directors at NIH were interested in focusing on genes relevant to their own institutes. According to Wyngaarden, scientists, cell biologists and genetics researchers in particular from the academy and others with contracts with NIH were threatened by the idea of big science. These scientists expressed concerns that sequencing projects would replace investigator initiated proposals and provide competition for funding. However, Wyngaarden reassured those with concerns that their jobs were secure.[6] Since there was a reluctance to change the NIH culture, it took time before acceptance of a large scale project with NIH as the lead agency occurred.

ACT II: NOT ON MY WATCH

In 1985, Wyngaarden secured the first NIH budget for human genomic research and awarded grants to researchers. In 1986, Charles DeLisi started the first government genome program. Upon hearing of DOE's financial commitment to genomics, Wyngaarden reacted by saying, "Not on my watch, if there is a major mapping and sequencing project, NIH is going to be the major player."[7] In 1988, to act on his commitment, Wyngaarden organized a meeting in Reston, Virginia to lobby for NIH as the lead agency.

Wyngaarden began a search for a project leader with a reputation in the field to secure the lead agency status and funds needed to meet its goals. That search led to James Watson who shared a 1962 Nobel Prize for Physiology or Medicine. The former Harvard professor ran a molecular biology lab and served as the Director of Cold Spring Harbor Laboratory. In 1988, Wyngaarden appointed Watson as Director of the NIH Genome Office. In 1989, NIH and DOE formed a joint program, and Wyngaarden changed the NIH Genome Office to the National Center for Human Genome Research.

The Human Genome Project officially started in 1990 with an ambitious checklist of scientific goals. The main goal was to obtain the highest resolution physical map which includes the entire DNA sequence of the genome with 2005 as the goal for completion. This is the first step to locating where disease genes are located. Other goals included identifying and mapping genes, placing the findings in databases, and developing tools for analysis.

Over the next decade, Applied Biosystems made significant advances in DNA sequencing capabilities. In 1991, with improved technology, scientists could sequence 10,000 bases per day at about a $1 per base, according to Michael Hunkapiller of Applied Biosystems. In 1993, robotics increased sequencing capabilities to 500,000 bases daily with a cost of 10-15 cents per base.

NIH did not perform the actual sequencing. The formation of an International Human Genome Sequencing Consortium allowed collaboration and provided higher sequencing resolution through assigning chromosomes to a number of sequencing centers. Five major academic centers referred to as the G-5 labs performed most of the sequencing. The G-5 included Washington University, MIT's Whitehead Institute, the Baylor College of Medicine, the DOE's Joint Genome Institute, and the Sanger Centre in England. The Wellcome Trust, a London based philanthropic organization similar to the Howard Hughes Medical Institute in the United States, and the U.K. Medical Research Council were the major sources of funding for the Sanger Centre which opened in 1993. Sir John Sulston, who formerly headed the Salk Institute, served as its director.

Researchers in the public project used several steps to sequence the genome; cloning the genome using bacteria and yeast, then breaking the clones into usable pieces roughly 150,000 nucleotides long starting with fragments whose positions on the chromosome were known by markers, then sequencing 600-800 bases at a time — the maximum sequencing machines could work with. Finally, millions of overlapping fragments required reassembly.

As the sequencing progressed, the original concerns with who could best lead a large scale project had resurfaced. In 1992, Watson contested NIH's Director Bernadine Healy for her support of what he referred to as the Venter patents.[8] While a researcher at NIH, Craig Venter's lab investigated the fight-or-flight response. Adrenaline has two responses in heart cells; it increases the rate of beating and the force of contraction. There he identified the protein on the surface of heart cells that senses adrenaline.

Venter used expressed sequence tags (ESTs), 300-500 base pairs of expressed RNA as a probe to find full length genes. In 1991, NIH began applying for patents on the ESTs. The USPTO rejected the EST patents; however, Healy NIH appealed the decision and applied for additional patents. Amidst charges of subordination and financial conflicts of interest, Watson left in April 1992 and returned to Cold Spring Harbor.

Francis Collins, who had a medical background, replaced Watson as director in 1993. Over the next several years a series of technological hurdles emerged. At the time, none of the consortium sequencing centers had substantial experience on larger genomes. John Sulston of the Sanger Centre in the U.K. and Robert Waterston of Washington University were involved in sequencing the roundworm *C. elegans* with 97 million bases from 1992-1998. In 1998 the journal *Science* revealed none of the genome centers funded by NIH were meeting their goals.[9] In 1997, halfway through the scheduled fifteen year long-range plan, the public project's researchers had sequenced only 3 percent of the human genome. In response, The Wellcome Trust and NIH increased funding to hopefully put the project back on schedule.

The public consortium continued their efforts in making physical and genetic maps for the human genome and model species easily accessible through online databases. A meeting held in Bermuda in 1996 resulted in the international public consortium volunteering to release all sequence data within twenty-four hours. Known as the Bermuda Rules, this applied to all DNA sequences with more than two thousand base pairs. In 1999, in an international effort to identify single nucleotide polymorphisms (SNPs) throughout the human genome, the Wellcome Trust and a group of pharmaceutical companies established the SNP Consortium which placed known SNPs in a public database.

ACT III: CELERITY, SWIFT IN MOTION

J. Craig Venter, a California surfer turned genomics pioneer, majored in biochemistry at the University of California at San Diego. After military service and graduate school in pharmacology and physiology, he spent ten years as a biochemistry professor at the Medical School at SUNY Buffalo.

In 1992, Venter resigned from a research position after NIH repeatedly denied him funding for his outside the box thinking. After establishing new research goals, Venter became more suited for the private sector. Venter's transition to the private sector landed him at Human Genome Sciences (HGS) where he received a lucrative deal. SmithKline-Beecham contributed $125 million in return for 7 percent equity while Venter received 10 percent equity in the company. With $85 million backing from venture capitalists, HGS formed a separate non-profit, The Institute for Genome Research (TIGR). SmithKline-Beecham would receive commercial rights to the genetic information and HGS handled marketing and licensing on sequence data from TIGR.

Venter left Human Genome Sciences in 1997 and formed Celera Genomics. Perkins-Elmer's Applied Biosystems division provided backing to Celera, which in turn received 80 percent of its profits, while chief scientist Venter received 10 percent. The name Celera was chosen for a purpose, which would serve notice to its competitor. It is derived from celerity which means swift and rapid in motion or action.

After observing the public consortium floundering, Venter would startle the science community. An article written by Nicholas Wade in the May 10, 1998 edition of the *New York Times* made public Celera Genomics' intentions of competing with the public sequencing project. Moreover, Celera's chief scientist declared he could accomplish the task at one-tenth of the cost with a $300 million budget. Based on the *Drosophila* sequencing project, the Celera group estimated it should take three years to sequence the human genome, a fifth of the time of the public project. As a result, Collins and former NIH director Harold Varmus were concerned that Congress might re-evaluate their funding.

In a talk at Duke University in April, 2004, Sir John Sulston recalled reading the story and was jolted by the news. He said the journals *Science* or *Nature* were usually the source of his professional enlightenment, and that the news was startling and humorous to those in the public consortium. In the words of James Shreeve, author of *The Genome War*, "It was the exceptional hubris of the plan that riled them."[10]

Harvard Professor Walter Gilbert once had similar commercial plans. In 1987, he resigned from a government post with the intentions of sequencing and patenting parts of the human genome. Gilbert's plan failed because he was unable to secure financing for a business venture. In contrast, Venter had a well developed business plan and partnership with Michael Hunkapiller, head of Perkin Elmer's Applied Biosystems division. Venter purchased an intimidating 230 state-of-the-art sequencing machines at $300,000 each enabling Celera to sequence hundreds of millions of bases in several months.[11] The race to sequence the human genome soon began.

Beside making business connections, Venter performed several small sequencing projects of his own. In 1995, Venter successfully sequenced the first genome of a free-living organism. Hamilton Smith had approached Venter about sequencing the 1,830,137 base-pair single celled bacteria *Haemophilus influenzae* genome by whole genome shotgun sequencing. Only viral genomes were sequenced before, which are considerably smaller. Frederick Sanger had sequenced the bacteriophage phi-X174 in 1975.

Gene Myers and J. Webers had discussed and proposed whole genome shotgun sequencing as an alternative to BAC to BAC for sequencing the human genome. Using this method, each chromosome is broken into large pieces and inserted into a bacterial artificial chromosome (BAC). The BAC is called a vector and the combination of the chromosomal DNA and the bacterial DNA is called a recombinant DNA molecule. The BACs are introduced into bacterial cells and replicate each time the cell divides.

Map-based clone-by-clone sequencing does not require the computer power that the whole shotgun method does. Rather than relying on maps using the BAC to BAC method, the shotgun method first breaks the genome into hundreds of thousands of pieces and relies on computer algorithms to reassemble them.

Members in the public consortium were skeptical of using shotgun sequencing for the human genome. Skeptics of the shotgun method perceived repeat sequences and the reassembly of fragments as serious technological hurdles. What Venter's critics considered technological hurdles, he considered challenges. Venter was confident that the shotgun strategy would succeed on humans as well. To reassemble the sequenced fragments, Venter hired top algorithm scientists and evaluated the top computer manufacturers in the world. The nucleotide sequences were successfully reassembled into the proper order using the whole genome assembly algorithm and the compartmentalized shotgun assembly.

Venter understood that with more complex genomes such as humans, maps with markers are critical for reassembling scaffolds. Sequenced tag sites (STSs) are unique stretches of DNA 200-500 nucleotides in length used in physical maps to assign specific places on chromosomes. Using polymerase chain reaction (PCR), Venter was able to amplify the STS markers. With the partial sequences as markers it is possible to pinpoint the gene's position on the chromosome.

Replicated DNA code is difficult to clone and presented a challenge for Celera. An innovative method devised by Hamilton Smith and Robert Holt called paired-end sequencing sequenced both ends of each cloned fragment of DNA, since both ends are unique. During the sequencing process, the clones were bar coded so that one could identify when the same clone was read from both ends, making reassembly easier.[12]

Celera used two types of data: genomes from five human donors, and cloned sequences downloaded from GenBank which were broken into fragments 2,000 to 50,000 nucleotides long. Using the public project's shredded sequences required reassembly, so they did not provide a significant time savings as the public consortium claims.

In spite of the technological challenges Celera finished a draft sequence in 2000. Amazingly, due to a boosted effort, the public project also had a draft sequence. Under pressure, James Kent, a computer programming graduate student at the University of California at Santa Cruz, pieced the public project's accumulated data together in four weeks after writing GigAssembler on multiple personal computers. Both drafts had over 100,000 gaps.

The competing public and private labs resulted in the Human Genome Project's completion ahead of schedule. While the draft sequence has 90 percent accuracy, a final draft, the gold standard, required 1 error in 10,000 bases or 99.99 percent accuracy. Closing these gaps meant more sub-clones from each cloned segment of DNA to fill the missing gaps. This required sequencing numerous times.

President Clinton chose Ari Patrinos of the DOE Joint Genome Institute, who held the infamous pizza-and-beer meetings at his home, to serve as a mediator between Venter and Collins for the public announcement. Venter and Collins eventually agreed to coordinate their announcements at a joint ceremony at the White House on June 26, 2001.

Feuding between the two projects extended into publishing their results. Bob Waterston, Maynard Olson of Washington University, Sulston, Eric Lander of MIT, Watson, and Collins of the public project threatened not to submit any future work to *Science* if they published Celera's work.[13] In February of 2001, *Science* published Celera's draft sequence, and *Nature* published the public consortium's draft sequence.

When chronicling the Human Genome Project, in many cases the science and innovative technological methods were lost in the shuffle. The media chose to focus more on the conflicting philosophies of the private and public projects. Backed by private funding, Venter was accountable to stockholders and management which based decisions on economics, not politics, which led to the necessary innovations.[14] Consequently, Venter's outside-the-box thinking led to using a number of innovative techniques including paired end sequencing, ESTs, whole shotgun sequencing, and the bar coding of repeat sequence which accelerated the sequencing of the human genome.

According to synthetic biologist Rob Carlson, the rapid increase in sequencing technology capability is the primary reason that the private effort by Celera was able to sequence the human genome so quickly.[15] The Applied Biosystems PRISM 3700 sequencers were less labor intensive and much more efficient. But, Venter is also responsible for obtaining the necessary computer technology necessary for reassembly. He had to acquire a fast computer hardware system that could perform using Celera's sequencing software. The possible candidates were narrowed down to Compaq and IBM. Compaq and its Alpha chip, the most powerful chip available at the time, won a multimillion dollar contract with Celera.[16]

In defense of the public project, Lander claims the project's timelines are misleading since the time consuming preliminary groundwork work of the public project is not reflected. According to Sulston, the cost of the Human Genome Project funded far more than sequencing.[17] The costs also included developing new technologies, maintaining databases, bioinformatics, closing the gaps, and the study of ethical, legal and social impact (ELSI) of genomics. Watson requested that 3 percent of federal budget for the HGP go towards proactively addressing the policy issues that would inevitably occur as a result of genomic research.

DEFINING GENOMICS

Defining the modern concept of genomics is easier said than done. Even categorizing genomics is not an easy task. Dictionaries define and journalists refer to genomics as a scientific discipline, an era, and a revolution.

GENOMICS, AN ERA?

In 2004 I sat in a as a guest in the first genomics class taught at Duke University. After a lecture, I asked the professor what the term "post-genomics" he spoke of means. At the time I thought my question was naive and was too embarrassed to ask him during class. But, it turns out it is not to be an easy question to answer. He replied, "The post-genomics era will entail sorting out what to do with the information obtained from the Human Genome Project." Post-genomics is sometimes more specifically referred to as the increasing emphasis on functional genomics or determining gene function.

In a special report in *The Economist*, Geoffrey Carr predicts that science historians will divide biology into the pre and post-genomics eras.[18] Not having fully grasped what exactly what the genomics era is, further research led to an article in *Nature Genetics* which states, "The genomics era is generally regarded to have started on July 28, 1995, with the publication of the genome of the bacterium *Haemophilus influenzae*."[19] But, this further added to my confusion since the April 19, 1981 issue of *Nature* reported that Frederick Sanger and colleagues sequenced the human mitochondrial genome in 1981.[20]

Francis Collins classifies proteomics as a subset of genomics, and defines genomics as more than sequencing genomes and expects it will remain ongoing for decades to come.

> Collins began a presentation by taking issue with the term post-genomics era. He queries whether this means that from the beginning of the universe until 2001 we were in the pre-genome era, and then suddenly we moved into the past-genomic era leading one to wonder what happened to the genomics era. Collins suggested that it was presumptuous to say that the Human Genome Project is already behind us.[21]

A genomics website helpful with definitions of various –omes gives a definition that places post-genomics in historical perspective.

> With an increasing number of organisms for which we have complete genomes we are beginning to see glimpses of the power of having fully mapped sequences. Still, in most contexts talking about being post-genomic seems a little premature. Post-Mendelian seems more accurate as we move from an era in which genetics has been rooted in monogenic diseases with high penetrance to a greater awareness of polygenic diseases and traits often with relatively low penetrance.[22]

I would argue post-Modern Synthesis or post-genetic program, the gene centered view to understanding diseases and evolution; however, I found a relationship that I feel comfortable with.

GENOMICS. A DISCIPLINE?

The American Heritage Dictionary defines genomics as "the study of" all of the nucleotide sequences, including structural genes, regulatory sequences, and non-coding DNA segments in the chromosomes of an organism. This denotes a discipline approach; however, it does not mention the study of gene products (RNA and proteins) or DNA interactions with other molecules which is a substantial portion of current genomics research.

Today, professors and researchers that specialize in the study of DNA may have thirty years of research experience, but chances are great they have never taken a genomics course. In most disciplines, it is customary for professors to have several degrees by taking numerous courses in their respective fields.

Typically academic disciplines such as genetics, psychology, and history professors are housed under one roof in a department setting. At MIT, Eric Lander heads the Whitehead Institute, and at Duke University, Huntington Willard heads the Institute for Genomic Sciences & Policy which are their respective genomic research institutes. "The institute status bridges disciplines" according to Willard. When asked if his research institute will become a department or school in the future, Willard responded, "When biochemistry achieved department status a good portion of the biology was lost, because a school or department status does not fully preserve the linkages between academic disciplines."

Most genomics research involves collaborations between disciplines including biology and computer science departments, and medical schools which are spread across campuses. So, genomics is not a discipline in the sense that psychology or history is, rather researchers from different disciplines provide an interdisciplinary approach to understanding puzzles in biology and medicine.

In the 1940s and 1950s, numerous researchers made award winning discoveries related to DNA. These researchers were classified as molecular biologists and may or may not have ever heard the word genomics. Molecular biology is defined as:

> The branch of biology that deals with the nature of biological phenomena at the molecular level through the study of DNA, RNA, proteins, and macromolecules involved in genetic information and cell function, characteristically making use of

advanced tools and techniques of separation, manipulation, imaging, and analysis.[23]

The *Merriam-Webster's Medical Dictionary* provides a more realistic definition of genomics:

> A branch of biotechnology concerned with applying the techniques of genetics and molecular biology to the genetic mapping and DNA sequencing of sets of genes or the complete genomes of selected organisms using high-speed methods, with organizing the results in databases, and with applications of the data in medicine or biology.

GENOMICS, A REVOLUTION?

In a little more than 130 years with biology as an academic discipline, scientists sequenced and mapped the human genome. The sequencing and mapping the human genome have led some to refer to a genomics as a revolution. Revolutions are distinctly different than eras. An era is a period or span of time marked from a fixed point or event, whereas a scientific revolution is a drastic change in ways of thinking or behaving. How does placing genetic markers and having roughly 3 billion letters of code in a cyberspace database warrant a revolution?

Galileo used a telescope to advance the understanding that the earth is rotating around the sun. Using this as an analogy, genomics is the telescope. Genomics has benefited from advances in biological equipment manufacturing and computing, which are industrial revolutions. These new tools and technologies have revolutionized the way scientists perform biological and medical research, and analyze data. In addition, interpreting genomic data has provided a paradigm shift in how scientists understand biology, a scientific or Kuhnian revolution. This, in turn, has an impact on experimental design.

EXPERIMENTAL DESIGN

Leading up to the Human Genome Project, scientists estimated the number of human protein coding genes was over 100,000 equaling the number of proteins. One of the most startling discoveries of the HGP was that a very small percentage of the roughly three billion nucleotide bases in the human genome encode instructions for the synthesis of proteins. These protein-coding DNA sequences referred to as exonic regions or exons comprise an estimated 2-3 percent of the human genome.

Post-genomic findings such as this have led to a shift from a molecular to systems level thinking. Systems biology requires a combination of approaches to better understand networks involved in biological systems. The systems level approach has developed into branches referred to as the (–omics).

To better understand complex phenotypes through gene and environmental interactions, researchers use (–omic) approaches to compare profiles under normal conditions and in response to stresses such as disease, changes in diet, and toxic exposure. Using (–omic) approaches, scientists can observe changes in proteins, transcripts, genes, expressed genes, and metabolite profiles in biological samples. After researchers generate data at many levels, they can place it in (–ome) databases accessible to other researchers.

A genome includes the complete set of genes in an organism; however, not all of those genes are expressed in a cell at any given time. For example, the cells in your big toes, kidneys, and eyeballs all have the same genome, yet their appearance is very different. This is due to differences in gene expression, which genes are turned on or off. Humans have an estimated 220 types of specialized cells, so what is an intron (non-protein coding DNA) and an exon is a matter of context. The genes involved with coding for eye color are exons in an eye cell; however, the same genes are introns in a kidney or toe cell.

In the past, analysis and gene expression was conducted on one or a few genes at a time. Analyzing many genes simultaneously required a new technology. Microarray technology has enabled researchers to decipher the phenome, the actual genes that are expressed in specialized cells. Using microarrays, researchers can determine which genes are expressed in a breast cancer tissue sample. Microarrays are also used to compare normal genes to diseased genes and the changes in gene expression over time under different environmental conditions.

Proteins and metabolites in bodily fluids interacting in networks play key roles in cellular function and regulating cellular processes. The complete set of proteins is the proteome. Besides identifying proteins, proteomics investigates the relationship between protein structure and function, protein interactions, and post-translational modifications. Researchers are attempting to identify key metabolites and investigate how they change under different circumstances. The complete set of metabolites is the metabolome.

Today, researchers are attempting to understand the genetic and environmental factors contributing to complex diseases. Genomic tools and technologies provide additional information to physicians for better diagnosis, earlier interventions, and more effective therapies for treating complex diseases. Researchers can now integrate genomics with traditional clinical testing, genetics, and epidemiology studies (see Table 1.1).

Table 1.1 Pre and post-genomic methods for understanding complex phenotypes

Traditional Medicine	Genomics
• clinical: blood work, pathology, lymph node status, X-rays, MRI imaging, CAT & PET scans • genetics: pedigree analysis, quantitative genetics, knock-outs, knock-ins, crosses • epidemiology: lifestyle, mental health, socioeconomics	• -omes & -omics databases: genome, transcriptome, proteome, metabalome, epigenome, phenome, exome, etc. • SNPs and haplotype mapping • the candidate gene approach • microarrays & gene expression • model organisms & comparative genomics

THE CANDIDATE GENE APPROACH

In the 1980s, the labs of Lap-Chee Tsui and Francis Collins used linkage analysis to find the gene associated with cystic fibrosis. Prior to the Human Genome Project, finding disease related genes was extremely expensive and time consuming. The researchers used prototype DNA sequencers to sequence DNA fragments possibly containing the cystic fibrosis gene. Researchers cloned several areas of the genome with linkage and compared them a physical map (see Figure 1.1).

Tsui and Collins understood that a piece of DNA close to the cystic fibrosis gene is rarely separated from the gene through recombination. Subsequently, they studied roughly 50 families that carry the disease and hundreds of DNA markers from blood samples to find markers — DNA sequences with a known location — that are inherited together to narrow down the genes position on a suspected chromosome. They discovered several markers, restriction fragment length polymorphisms (RFLPs), that migrated together during recombination.

Using computer programs, they constructed a linkage map based on genetic and physical markers with the disease phenotype. In 1989, Collins

linkage analysis → gene mapping → **physical mapping** → **gene cloning** → ID mutation

Figure 1.1 Pre-Human Genome Project method for determining disease related genes[25]

and Tsui discovered the a disease related mutation on the cystic fibrosis gene located on Chromosome 7.[24]

Genomics has revolutionized the way researchers go about finding disease related genes. By comparing sequenced genomes of individuals with and without diseases, researchers can compare the differences. Comparing genomes allows the candidate gene approach to finding disease genes (see Figure 1.2). Today, researchers can pinpoint the disease locus and search available online databases for candidate genes near designated markers. The Human Genome Project eliminated the cloning and physical mapping steps.

linkage analysis → gene mapping → genome database → ID mutation

Figure 1.2 Post-Human Genome Project method for determining disease related genes: the candidate gene approach[26]

IN SILICO ANALYSIS

The -omes & -omics approach to studying human diseases has provided researchers with extremely large amounts of data to work with at numerous hierarchal levels. This is mainly due to the commercialization of the –omics approaches. Much of this data is placed in open source databases such as PubMed, a service of the U.S. National Library of Medicine, and GenBank which is managed by NIH.[27]

These databases also include a number of model organisms which are useful for understanding evolutionary relationships and in medical research. Although the mouse diverged from humans evolutionarily approximately 60 million years ago, it has conserved sequences in most biological systems. Comparative genomics utilizes homologs, DNA sequences related by common ancestry, to assist with determining functions of human genes. The Mouse Sequencing Consortium has provided a sequencing database to identify the function of homologous human genes. Using comparative genomics, researchers discovered that 99 percent of mouse genes have homologs in man. Conservation means functionality. Of these DNA sequences, 96 percent have synteny; the same relative location on the chromosome. However, the actual protein similarity between mice and humans is less than 99 percent.

In addition to the mouse, which is used in circumstances when it is unethical to use humans, researchers now have the ability to access the genomes of the model organisms. These model organisms are important to researchers for comparing exonic (non-coding DNA) and intronic (coding

DNA) regions, and to identify regulatory genes. Among the commonly used model organisms are yeast, *E. coli* bacteria, the roundworm, and the fruit fly. Thomas Hunt Morgan used the fruit fly in the early 1900s. In 1974, Sydney Brenner first used the roundworm *C. elegans* because it is one of the simplest organisms with a nervous system.

Table 1.2 Comparison of number of genes in humans and model organisms

Organism	Date sequenced	# of genes
Saccharomyces cerevisiae (yeast)	1996	6,034
Mus musculus (mouse)	1996	~22,500
Escherichia coli (bacterium)	1997	~4,500
Caenorabditis elegans (roundworm)	1998	19,099
Drosophila melanogaster (fruit fly)	2000	~13,600
Homo sapiens	2003	~24,000
Pan troglodytes (chimpanzee)	2005	~24,000

After researchers obtain vast amounts and levels of biological data, there remains an equally important task of managing and analyzing the data. Bioinformatics, using computing and statistical techniques, enables researchers to manage the vast amounts of data generated by the -omic & -omics approaches.

In 1990, David Lipman and Eugene Myers of the National Center for Biotechnology Information created the BLAST algorithm to analyze data. The BLAST software program compares the functional and evolutionary relationships making it easier to search for homologous nucleotide and protein sequences. Bioinformatics enables researchers to turn data into useful information and ultimately knowledge of how biological systems work (see Figure 1.3).

research data → bioinformatics → information → knowledge

Figure 1.3 Management and analysis of research data

CHAPTER 2

Mining Genomes

The concept of a human genome is somewhat a misnomer; rather it is a reference genome. With Venter's estimation that human genomes are 99.5 percent similar with roughly 3 billion base pairs in each human genome, this means approximately fifteen million nucleotide differences exist between any two people. Polymorphisms that occur in different frequencies around the world are used to sort people into genetically different groups.

Austrian pathologist Karl Landsteiner was the first to discover a polymorphism, a nucleotide mutation, in humans. In 1901, he discovered proteins on the surface of blood cells that led to the development the ABO blood typing system. During blood transfusions, matching blood types is necessary due to reactions between antibodies of the recipient and antigens found on the surface of red blood cells of the donor. Red blood cells carry oxygen in the blood. In 1958, Jean Dausset discovered the antigens of the ABO and Rh systems which are located on the surface of white blood cells which are import for immunity.

The human leukocyte antigen (HLA) complex located on chromosome 6 contains over 200 genes. Among these are over 40 genes that encode leukocyte antigens which are cell surface proteins essential for immune response.[1] The HLA complex is the second most polymorphic set of genes found in humans after variable number of tandem repeats (VNTRs).[2] In 1984, Alex Jeffreys accidentally discovered VNTRs which are located near genes. These short repeats are distinct in populations and are useful for forensic identification. One common VNTR is the DNA sequence "CACACA..."

Natural selection has created the polymorphisms that make the ABO blood typing system and HLA complex unique among human populations to fight diseases. Similarly, researchers want to mine diverse human genomes and provide specific medical treatments to individuals and populations that nature has yet to create.

To identify genes and mutations that make humans susceptible to most diseases, it is important that researchers have DNA samples and a large database of genomes to analyze. As the price of sequencing genomes continues to decrease, researchers will have a more realistic opportunity to use population studies and develop databases to receive the maximal benefit from genomic biomarkers.

Two current independent projects aim to mine gene variants and correlate them to diseases and human differences. George Church of Harvard Medical School started the Personal Genome Project in hopes of integrating 100,000 individual genomes with their medical history. Among the projects goals are to develop tools to interpret genomic data.[3] The 1000 Genomes Project is an international effort to develop a detailed catalog of human genetic variations through association studies to discover SNPs, insertions or deletions (indels), and copy number variations (CNVs) which are large regions ranging from ten thousand to five million nucleotides that are deleted or amplified.[4] These ambitious long-term projects will probably have to wait for the magical $1000 cost for sequencing human genomes to delivery any groundbreaking results.

Scientists have discovered two types of gene variants related to disease. One type is rare mutations that lead to severe diseases. In the 1980s, researchers discovered mutations linked to cystic fibrosis, Huntington's disease, and breast cancer using linkage analysis. However, these types of diseases are rare. The second type of gene variant is more common and leads to complex diseases. These biomarkers are more difficult to find since most complex diseases do not follow Mendelian inheritance. Most complex diseases have multiple origins linked to multiple sets of genes and environmental factors.

The Human Genome Project inspired biotechnology companies to search for biomarkers associated with complex diseases including asthma, diabetes, heart disease, cancers, Alzheimer's disease, arthritis, obesity, and behavioral disorders. To locate susceptibility genes, researchers can use individual genomes in two ways. They can compare patients samples with know variants in regions cataloged in databases which save researchers the time and expense of sequencing genomes. Also, researchers can perform whole genome association studies to determine gene variants linked to diseases.

Genetic maps are useful for identifying disease related genes. Genetic mapping is based on the frequency that genes are inherited together during meiosis. Bioinformatics has revealed that the closer genes are together, the greater the odds that during recombination that these genes and SNPs of interest are inherited together. Using linkage disequilibrium, the occurrence of combinations of alleles in a population more or less often than expected, researchers can locate candidate genes associated with diseases. These blocks of inherited SNPS are called haplotypes and are distinct in groups of people. This finding contradicts Mendel's Law of Independent Assortment that states that hereditary units now referred to as genes segregate independently and have an equal chance of appearing in offspring.

Almost half of European-American nucleotide sequences are contained in haplotype blocks, compared to only a fourth of African American sequences.[5] Scientists believe this is because of a population bottleneck occurring after the first humans migrated out of Africa. A significant percentage of the population was killed or otherwise prevented from reproducing, reducing the gene pool. Consequently, researchers need more SNPs for studies on older lineages.

DEFINING GENETIC PROPERTY RIGHTS

Although the primary use for the DNA samples is for medical and scientific reasons, third parties can potentially misuse the information encoded in DNA. This is a concern for activists who have subsequently lobbied for legislation and oversight on issues regarding the potential misuse. The activist's primary public policy issues relate to informed consent, confidentiality, discrimination, and the respect for the beliefs of indigenous peoples.

An analysis of these bioethical issues reveals that they are not so much problems with biotechnology itself, rather oversight and regulations have led to government failures. These failures, in turn, require further analysis and government intervention. This raises the question, in a democratic, free market, and capitalist society, how the government should intervene to ensure these genetic property rights.

One approach is to seek guidance from the Constitution. Commenting on the relationship between bioethics and the Constitution, Justice Antonin Scalia notes the tools for this job are not to be found in the lawyer — and hence not in the judge workbox.[6] Since the Constitution is silent on bioethics, these issues are left to the three branches of government to resolve.

For a free market system to work efficiently, it is necessary to clearly define and enforce property rights and to protect citizens. In 2005, in response to public policy issues involving genetic property rights, the Council for Responsible Genetics (CRG) proposed a Genetic Bill of Rights (see Table 2.1). These public policy issues related to biotechnology and genetic property rights do not necessarily divide across Republican and Democratic party lines. Rather, they are part of a broader culture war based on values. Since the CRG's proposal is from a civil society perspective, some articles proposed by CRG are objectionable to libertarians.

Table 2.1 A Genetics Bill of Rights proposed by the Council for Responsible Genetics[7]

Articles
1. the preservation of the earth's biological and genetic diversity
2. no patented DNA
3. no genetically engineered food supply
4. protection of indigenous peoples from biopiracy and to preserve traditional knowledge
5. protection from toxins to humans and our offspring
6. protection against negative eugenics including forced sterilization and screening for genetic manipulation in embryos
7. informed consent on use and storage of biological samples
8. protection against genetic discrimination
9. the right to a DNA testing for defense in civil and criminal cases
10. the right to be born without genetic manipulation

WHY ELSI?

As part of the government role of protecting citizens, the Ethical, Legal, and Social Issues (ELSI) program is now required for government funded projects using human samples. It has also become the largest bioethics program and a model for similar programs around the world.

Civil libertarians are passionate activists for ethical legislation and go into detail on the many possible injustices that may occur. However, they provide little explanation on the historical precedent that would justify the action of why citizens need protection and who those submitting DNA samples actually need protection from. The implications of writings and legislation proposed by civil libertarians are geared towards private companies. But, is public policy geared towards private companies actually warranted based on historical precedent?

FEEBLEMINDED GENES: *BUCK V. BELL* (1927)

In the 1883, Francis Galton coined the word eugenics, meaning well-born. Galton established the Laboratory for National Eugenics at the University College of London and formed the first genetics department. He also helped develop the field of biometrics, applying statistics to investigate the linkage of heritability with traits such as character, talent, and intelligence.

Galton's followers believed that mental deficiencies including antisocial behavior and feeble-mindedness as well as successful traits are heritable. As population genetics research developed, so did a eugenics movement. Negative eugenics is the elimination of undesirable traits. Professional mental health professionals, physicians, and scientists were the ones who understood genetics and promoted social improvement by reducing crime, prostitution, slums, and alcoholism through eliminating undesirable traits.

In 1904, while teaching at Harvard, Charles Davenport convinced The Carnegie Institution to spend ten million dollars to establish a Station for Experimental Evolution at Cold Spring Harbor. Through this research, researchers acquired three generations of family records for the feebleminded. Harry Laughlin, the director of the eugenics office at Cold Spring Harbor, developed a sterilization project as a solution to feeblemindedness. The project included voluntary and in certain cases compulsory sterilization of those with learning difficulties and mental deficiencies, including the homeless and criminals. Indiana passed the first compulsory sterilization law in 1907.

Laughlin designed a model law that institutionalized the feebleminded in the state's best interest and prevented the prosecution of physicians. In some cases the patients were sterilized by cutting the fallopian tubes to prevent reproduction. In 1924, The Virginia General Assembly passed The Eugenical Sterilization Act passed, and 17 year old Carrie Buck of Charlottesville, Virginia became a guinea pig to test the law. Carrie Buck's foster parents committed her to the Virginia Colony for the Epileptic and Feebleminded which opened in 1910 after she gave birth to an illegitimate child. Carrie's mother was earlier committed to the same asylum. Officials of the Virginia Colony were convinced she had inherited the traits of feeblemindedness and sexual promiscuity. Furthermore, the officials believed Carrie's daughter Vivian was likely to become an imbecile like her mother and grandmother.

The U.S. Supreme Court Case *Buck v. Bell* (1927) upheld Virginia's Eugenical Sterilization Act by a vote of 8-1 allowing involuntary sterilizing of epileptic and feebleminded inmates at the Virginia Colony for Epileptics and FeebleMinded in Lynchburg, Virginia. In the trial, attorney Irving Whitehead, who was previously legal counsel to the Virginia Colony, provided a weak defense of Buck. Whitehead did not call witnesses to challenge charges about her mental health, and the fact that a relative of Carrie's foster parents had raped her was never raised in the court proceedings.[8] Laughlin never met any members of the Buck family, but sent sworn testimony that the Buck family was feebleminded.

In the majority opinion for *Buck v. Bell*, Justice Oliver Wendell Holmes wrote, "Society can prevent those who are manifestly unfit from continuing their kind. Three generations of imbeciles are enough." Justice Holmes used *Jacobson v. Massachusetts* (1904), which upheld a Massachusetts law that required the vaccination of school children against smallpox, in support of the Court's decision.

As a result of *Buck v. Bell*, sterilization rates in the United States climbed from 1927 until *Skinner v. Oklahoma* (1942). Under a 1935 Oklahoma Act, involuntary sterilization became compulsory for individuals convicted of three felonies. In the case of Skinner who was convicted of stealing chickens and armed robbery numerous times, the Supreme Court held that the Oklahoma law violated the Equal Protection Clause of the Fourteenth Amendment since the law excluded white collar crimes including embezzlement and tax evasion.

In the 1930s, even the well-born in the United States found themselves socially outcast with the severe economic depression casting doubts on the role of genetics.[9] However, the eugenics movement would continue into the 1950s and led to more than 60,000 court ordered sterilizations carried out in over thirty states and two Canadian provinces. Then, each state involved in involuntary sterilization passed legislation banning the practice.

The eugenics movement spread to Latin America, Asia, and Europe. Although eugenics is usually associated with Germany and the Nazi Third Reich's programs that targeted Jews, gypsies, and homosexuals that subsequently led to the Holocaust, in reality, the German eugenics programs were based on practices in the United States.

The Third Reich adopted sterilization laws in 1933 and negative eugenics legislation in three 1935 Nuremberg Laws. These laws applied to the handicapped and those suffering from diseases and behavioral disorders. In Germany, an estimated 400,000 people were sterilized, roughly 5 percent of the population. Originally, the Third Reich sterilized males by vasectomy and females by tubular ligation. Since these procedures are time consuming, the Third Reich later castrated males and exposed women to X-rays.

The accumulation of people to house and feed eventually led to the passage of euthanasia laws. The Third Reich was also interested in the relationship between brain structure and mental illnesses. A number of sterilized victims were later subjected to inhumane genetic experiments. The Kaiser Wilhelm Institute for Brain Research examined brains of the euthanized persons.

THE U.S. PUBLIC HEALTH SERVICE (1932-1972)

From 1932 and 1972, the U.S. Public Health Service conducted a study on 399 African American sharecroppers with syphilis for research on the disease's natural progression. The test subjects received free meals, medical exams, and burial costs in exchange for their participation. Now known as the Tuskegee syphilis study, the participants were not aware they had syphilis, nor did they receive treatment. In 1972, a source brought the study to the public's attention with coverage in major national newspapers. This study led to the 1979 Belmont Report which set guidelines for the protection of human subjects. As a result Institutional Review Boards now review the procedures for studies on human subjects.

In 2010, while researching a book on the Tuskegee study, Susan Reverby of Wellesley College discovered archived notes at the University of Pittsburgh revealing the project was more widespread. Dr. John Cutler, a professor at the University of Pittsburgh, was involved in the Tuskegee study as well as another study in cooperation with the Guatemala government. From 1946 to 1948, the U.S. Public Health Service Sexually Transmitted Disease Inoculation Study intentionally infected Guatemalans with sexually transmitted diseases to determine the effectiveness of penicillin. At the time of the study, prostitution was legal in Guatemala. A total of 772 people including female sex workers, prisoners, and mental patients were infected with gonorrhea, 696 with syphilis, and 142 with chancres. NIH reports indicate dozens of other similar studies have occurred.[10]

WAFFLING ON SAVING THE SNAIL DARTER: *TVA V. HILL* (1978)

The Department of the Interior established The Tennessee Valley Authority (TVA) during the Great Depression to bring electricity to rural parts of the South. In 1967, The TVA launched The Tellico Dam and River Project to provide hydroelectric power, a reservoir, shoreline development, recreation, and flood control.

In 1973, University of Tennessee ichthyologist (a zoologist who studies fish) Dr. David Etnier discovered an unknown variety of perch, *Percina tanasi*, also known as the snail darter, living in the Little Tennessee River. The timing was unfortunate for the Tellico Dam Project since the environmental movement was gaining momentum. In 1974, the Fish and Wildlife Service, which also falls under the Department of the Interior, established the snail darter as a separate species and placed it on the endangered species list. The standing in court of an obscure Southern fish species would become the subject of a legal battle when placed against the economic development of its habitat.

In 1976, University of Tennessee law student Hiram Hill and his environmental law professor Zygmunt Plater initiated a lawsuit to prevent completion of the $116 million dollar project based on Sections 7 and 11 of the Endangered Species Act (1973) which address the preservation of the earth's biological diversity. The District Court agreed that the project would destroy the snail darter's habitat. Once a federal project is shown to jeopardize an endangered species a court must issue an injunction that will halt development of a project. However, a judicial ruling stated since roughly 80 percent of the project was completed, stopping the project would waste millions of taxpayer dollars. So, Congress continued to fund the project.

A Federal Appeals Court ruled that all construction activity must stop until Congress exempted the Tellico Dam Project from compliance with the Endangered Species Act or the snail darter was no longer in danger of extinction. In an attempt to preserve the snail darters, TVA transplanted the fish to another section of the Little Tennessee River, but could not demonstrate that the transplanted fish could successfully reproduce. A legislative committee denied the project an exemption.

In an attempt to resolve the impasse between the endangered species and the Tellico Dam Project, the TVA's next step was to petition the Supreme Court for review. In 1978, the Supreme Court granted the request. In testimony for *TVA v. Hill*, Professor Plater argued the TVA project violates the reasoning behind why Congress passed the Endangered Species Act to begin with: the need to avoid further reduction of national and worldwide wildlife resources.

In explaining the need for legislation, the Report of the House Committee on Merchant Marine and Fisheries on H.R. 37, a bill that became the template for the Endangered Species Act of 1973, stated:

From the most narrow possible point of view, it is in the best

interests of mankind to minimize the losses of genetic variations. The reason is simple: they are potential resources. They
are keys to puzzles which we cannot solve, and may provide
answers to questions which we have not yet learned to ask.

To take a homely, but apt, example: one of the critical chemicals in the regulation of ovulations in humans was found in
a common plant. Once discovered, and analyzed, humans
could duplicate it synthetically, but had it never existed — or
had it been driven out of existence before we knew its poten-
tialities — we would never have tried to synthesize it in the
first place.

Who knows, or can say, what potential cures for cancer or
other scourges, present or future, may lie locked up in the
structures of plants which may yet be undiscovered, much
less analyzed? …Sheer self-interest impels us to be cautious.[11]

The U.S. Supreme Court ruled in favor of the snail darter over the
completion of the Tellico Dam. The high court interpreted the Endangered Species Act as requiring saving the species whatever the cost. This
ruling established that animals and plants have standing in court. The
blue-haired, tennis shoe wearing little old ladies, aka conservationists and
environmental activists, who worked against the dam's development were
ready to declare a victory in the political process. However, the case was
not over.

The project was completed and the developers ultimately prevailed.
Congress decided to give more flexibility in decisions when balancing
economic interests with rare gene pools which could someday provide
medicinal value. In 1979, Senator Howard Baker and Representative John
Duncan placed a rider as part of the Energy and Water Development Appropriations Act of 1980 which specifically exempted the dam project
from the Endangered Species Act compliance.[12]

In the process of enforcing the laws enacted to protect property rights,
when politicians and bureaucrats acting on behalf of the public decide to
act in self interest by trading votes on another project, it results in a government failure.

THE DEMISE OF THE HUMAN GENOME DIVERSITY PROJECT

In 1991, a group of scientists led by Stanford's Luca Cavalli-Sforza initiated the Human Genome Diversity Project (HGDP) via a letter in the journal *Genomics*. The project proposed the Morrison Institute for Population and Resource Studies at Stanford University would collect blood and tissue samples, and cell lines from around the world with a representative database of human genetic diversity. For research purposes, scientists could grow the cell lines in the laboratory as needed.

The scientists' proposal stressed urgency, since a number of indigenous groups were vanishing. Before these populations were extinct it was important to preserve their rare gene pools. With DNA samples from over 700 communities around the world, the database would contain the sequences with the estimated 0.5 percent difference among humans. In 1988, privately supported funds launched the Human Genome Organization (HUGO) based at Cold Spring Harbor Laboratory which provided international scientists a clearinghouse for research data.

Researchers would enter the samples into a database for the purpose of benefiting humanity and other researchers. In 1991, the American government adopted the Common Rule, a policy adopted for the protection of human subjects that applies to 17 federal agencies regarding human testing and government funded research. Section 46.111 provides criteria for Institutional Review Board approval of research and section 46.116 provides general requirements for informed consent.[13]

Although the project claimed the study would provide medical applications unique to specific ethnic groups, it was controversial. Indigenous populations and activists groups were wary because of colonialist's past exploitation of humans and taking their land. With the belief in some African communities that researchers purposely spread AIDS in Africa possibly from an experimental polio vaccine tested in the Congo, there were also concerns of racism and the possible creation of biological weapons to target certain populations. In addition, the potential commercialization of the DNA made activists skeptical of these claims. The activists concerns were justified as the U. S. government subsequently filed for patents on DNA samples.

In 1993, Ron Brown, Clinton's Secretary of Commerce filed a patent claim on a cell line of a twenty-six year old Guyami woman from Panama. She had a unique virus that made her antibodies useful in AIDS and leukemia research. A protest by a number of groups including the Guyami people, advocates for indigenous populations including the NGO Rural Advancement Foundation International (RAFI) now known as ETC

Group, and the World Council of Indigenous Peoples (WCIP) led to the withdrawal of the patent application.

In 1994 the U.S. government filed for a patent on a cell line from the Hagahai people in Papua, New Guinea that carry the gene that predisposes humans to leukemia. Despite objections from activists, the United States Patent and Trademark Office granted the patent to the Department of Health and Human Services. The USPTO, which approved the patent is under the jurisdiction of the U.S. Department of Commerce. Under pressure, the Commerce Secretary Ron Brown responded to activists saying that under American laws what is patentable is not dependent on the source of the cells.[14] However, the Hagahai patent produced little financial rewards, so the rights to the patent were subsequently abandoned.[15]

Debra Harry, an activist and a member of The Indigenous Peoples Council on Biocolonialism, an advocacy group formed in 1993, called for a boycott of the project to the WCIP. The opposition Luca Cavalli-Sforza and his colleagues received from these groups led to the programs demise. In 1997, a report from the National Academy of Sciences recommended that the National Science Foundation and the NIH no longer fund the HGDP.

In a second attempt to make a family tree of human populations, researchers are taking a different approach. With backing from IBM and the Waitt Family Foundation, the National Geographic Society has revived the HGDP project. The Genographic Project also receives funding through $99.95 DNA-sampling kits from participants that have polymorphisms in their mitochondrial DNA or Y chromosomes analyzed.

This time, no patents or medical research will result from the collected DNA samples. The project only uses the samples for the limited scientific objectives, and the researchers do not develop cell lines from the DNA samples. Rather, the samples are deposited in regional biorepositories, analyzed, and then destroyed. The project's director, Spencer Wells, has independent committees, typically from local universities and governments, that approve research protocols. The participants in the project are volunteers that provide informed consent and may withdraw their consent at any time.[16] Rather than using DNA samples for profit, the Genographic Project established the Genographic Legacy Fund to provide financial support for the preservation and development of indigenous populations.

Another potential concern with analyzing DNA from indigenous populations is how these groups will respond to information that is contradictory to their traditional beliefs.

When Wells talked about his book *The Journey of Man* at Duke University in 2004, he told the audience of informing an Aboriginal leader that his ancestors are from East Africa. The leader absolutely discounted Wells' scientific explanation for Aboriginal ancestry. Indigenous groups have strong cultural beliefs about their origins and sense of self. Since indigenous peoples have their own stories passed on orally, they have the right to determine if the scientific explanation will complement their own world view or reject it. So, Wells diplomatically told the leader this is the modern scientific story, your songlines are yours.

MEDICAL RESEARCH BIOREPOSITORIES

Hospitals collect and destroy many biological samples. If those samples were kept and placed in biorepositories, researchers could later use them for detecting patterns of adverse drug reactions and to determine genetic relationships with diseases. In the past some hospitals were hesitant to use stored DNA samples for fear of law suits.

Genomic information is different from a physical examination or a blood test because biomarkers in individuals and populations may provide susceptibility to diseases. As a result, some people are unwilling to provide DNA samples for research because of legacy issues related to a lack of clearly defined genetic property rights. In disputes over commercialization rights related to DNA samples, district and state courts have ruled unpredictably in the absence of federal legislation.

According to Julia Giner of the Clinical Research Institute at Duke University hospitals will wait until safeguards are in place for possible misuse of lab samples in order to prevent litigation.

WASHINGTON UNIVERSITY V. CATALONA (2003)

Washington University has collected tissue samples from thousands of patients. Many patients intended for William Catalona, a prostate cancer specialist and developer of the PSA diagnostic test, as the recipient of the samples. The university's intentions were inconsistent with the research the donors had consented to. So, Catalona left Washington University for Northwesten University, after conflicts over access to the samples.

Over six thousand patients agreed to have their samples transferred to Northwestern University. However, Washington University refused to the request for transfer and subsequently made the samples anonymous by removing donor names from the DNA samples. So, who should have control of samples collected over decades and stored for research?

In 2003, Washington University sued Catalona for ownership of the samples. Even though the donors had given the samples to Catalona and had the right to withdraw from the university research, a judge ruled in favor of the Washington University. Judge Limbaugh wrote in his opinion that the utility of biorepositories is seriously threatened if samples were moved anytime a donor requested them. In addition, samples that are not traceable to a specific person are exempted from the right to withdraw.[17]

GREENBERG V. MIAMI CHILDREN'S HOSPITAL RESEARCH INSTITUTE (2003)

Daniel and Debbie Greenberg lost two children to Canavan disease which is characterized by seizures, a vegetative state, and typically fatal before four years of age. If both parents carry a copy of the Canavan gene, primarily found in Ashkenazi Jews, children have a 25 percent chance of inheriting the disease.

Dr. Reuben Matalon and Miami Children's Hospital received funding from the Canavan Foundation to identify the disease gene and to develop a screening test. The hospital and Dr. Matalon, without informed consent from the donors, obtained a patent on the gene in 1987 and charged a licensing fee for testing. The families donating tissue samples and organizations donating money alleged they were not aware of the commercial intentions. Rather they understood the diagnostic tests would remain in the public domain. In *Greenberg v. Miami Children's Hospital Research Institute* (2003), the Florida District Court dismissed the informed consent claim on the basis that research subjects could dictate how medical research progresses.[18]

MOORE V. REGENTS OF THE UNIVERSITY OF CALIFORNIA (1990)

After a diagnosis of hairy cell leukemia in 1976, John Moore became a patient of David Golde at the UCLA Medical Center. Golde advised Moore to have a splenectomy, the removal of his large spleen, to slow the progress of the disease. Some of Moore's blood products had potential commercial value since they may help provide cures for diseases. In 1981, the University of California filed for a patent on the "Mo" cell line, which was established using Moore's T-lymphocytes (a type of white blood cell). Moore's T-lymphocyte's overproduced a type of T-lymphokine which had potentially therapeutic properties. In 1984, Dr. Golde received patents on the cell line and a method for producing lymphokines from the cell line.

In *Moore v. Regents of the University of California*, Moore claimed his physician had failed to disclose his economic interests in the donated tissue and wanted a share of any royalties. In 1983, Moore had signed a consent form as a research study participant, but refused to grant the University of California any rights to cell lines developed from tissue samples. In 1990, the California Supreme Court recognized Moore's rights of informed consent, but denied his claims of property rights to the cell line. So, Moore had no rights to any profits.

The wording in the consent forms and state laws are important in litigation. For example, whether or not consent forms expressly allow or restrict use of donated tissue. In this particular case, California's statutory law limits the patient's control over excised cells.[19] Moore's conversion claim, the unauthorized right of ownership over goods belonging to another, failed, since Moore had no expectation of retaining possession of his tissues and showed no interference with his right of possession of his cells without title or possession. In addition, the court ruled the products derived from Moore's cells are distinct from his tissues.

HAVASUPAI TRIBE V. ARIZONA BOARD OF REGENTS (2008)

Today, roughly 600 Havasupai Indians live in Supai, a village floor reservation in the Grand Canyon. The tribe has lived here for over 800 years. Modernization has brought changes to their culture, including satellite dishes and an extraordinary rise in Type 2 diabetes over the last three generations.

John Martin, an anthropology professor at Arizona State University, spent a year with the tribe in 1963 for his doctoral dissertation, and developed a trusting relationship with the tribe. He later recruited two researchers to study the diabetes epidemic. In 1989, the researchers received grants for a study, and over 100 tribe members provided blood samples. During the course of their research, the researchers noticed in the tribe member's medical records a 7 percent incidence of schizophrenia. The frequency is 1 percent in the overall United States population. In 1990, the researchers received a grant to study schizophrenia within the tribe.

A tribe member attended a defense of doctoral dissertation relating to the study and observed not much was related to or helpful relating to diabetes. The research revealed the Havasupai do not share the gene variant with the Pima Tribe linked to diabetes, which indicates that any genetic links to the disease were acquired through a different mutation. Other than education through nutrition classes, the tribe felt the study's purpose was to further Western science rather than to preserve their culture.

A university investigation revealed two dozen published articles based on the blood samples. One article proposed the tribes inbreeding correlates to higher susceptibility of these diseases. Another article revealed the tribe originated from Asia, contradicting their traditional beliefs and identity. The tribe is raised to believe the Grand Canyon is the birthplace of all humans. Having Asian heritage would also threaten their sovereign right to the land.

In 2004 the tribe filed a lawsuit against Arizona State University over a disagreement over fully informed consent and the handling of DNA samples. Several samples were lost and two had the identities of the donors, which were not necessary for the agreed upon research and could potentially infringe on the donors' privacy. Although the tribe attempted to move the case from the state to the federal level, the federal court remanded the case back to a state court for lack of federal jurisdiction. After spending $1.7 million to defend the lawsuit, the testimony revealed conflicting information regarding consent of the samples.

In a settlement, Arizona State paid the tribe $700,000. The researchers also returned the remaining blood samples, which have a deep spiritual meaning to the tribe.[20] This case reveals the importance of the need for the Intituteional Review Board to review research protocols prior to a research project in order to provide a clear understanding of the information contained in DNA samples and the specifics of informed consent.[21]

DeCODE GENETICS

The ethnically homogeneous people of Iceland are descendants of a few Norwegian and British explorers dating back eleven-hundred years. Common ancestry and well kept genealogical records make Iceland an ideal location for genomics research. In 1996, the Icelandic government contracted the Icelandic company deCODE Genetics to place the health records of its roughly 260,000 citizens in a centralized database.

In 1998 the Icelandic Parliament passed the Health Sector Database Act which provided a license to place its citizen's health records in a national database. With a heated debate in the media, Icelanders were educated about the groundbreaking undertaking and 93 percent of the population supported it. For citizens to keep their records out of the database, the burden was placed on the individual to send a form to the Ministry of Health similar to the no-call list in the United States.[22]

Roche Pharmaceuticals, the major backer of the project, granted free drugs to Icelanders and received the rights to any diagnostic tools or drugs resulting from the research. In return, Kari Stefansson, deCODE Genetics' President, agreed to find genes for twelve diseases and received exclusive rights to the data.

It appears a win-win situation. However, at the time Iceland had no laws to protect its citizen's confidentiality. So, a controversy over presumed consent versus informed consent led to law-suits, requiring the Icelandic government to temporarily halt the project. In 2007, the Icelandic Supreme Court struck down the Icelandic Health Sector Database Act as unconstitutional. Consequently, it is no longer mandatory to send health records to the national database.

WHY GINA?

The recent explosion of sequenced DNA genomic information has also raised concerns of genetic discrimination. Without non-discrimination laws there is the potential that insurers, courts, schools, corporations, and the military who obtain genomic information could use it against individuals. Insurance rates could rise with the knowledge of an increased risk of a particular genetic disorder or future medical condition. To avoid genetic discrimination, genetic counselors and other health officials are obligated not to release information unless granted permission.

In the absence of federal legislation related to insurance discrimination, laws will vary from state to state. Persistent pressure from activists has led to a series of federal policies. In 1995, the U.S. Equal Employment Opportunity Commission modified the definition of disability. The rewording aims to protect individuals from discrimination based of genetic information relating to illness. In 1996, Congress passed the Employee Retirement and Income Security Act (ERISA) outlawing discrimination in group health insurance plans. The law takes current health status for a pre-existing condition into consideration, but not genetic information.

In 2000, President Clinton signed an executive order preventing genetic discrimination to federal job holders. The order also provides privacy protection on medical treatment and research.

Congress passed the Genetic Information Nondiscrimination Act (GINA) in 2008 so people can take diagnostic tests and participate in clinical trials without fear from individual health insurance or employment discrimination. GINA also prohibits the use of individual genetic information by employers for decisions in firing, hiring, job assignments, and promotions.

Since there is little evidence of employment discrimination in the private sector based on genetic information, activists in favor of free markets argue that GINA is a proactive solution in search of a problem. But, is this a moot point? Policy Analysts at the Competitive Enterprise Institute, a libertarian think tank, argue, "Because nearly all diseases with a genetic component can be prevented or treated with early detection, widespread genetic testing is far more likely to result in improved health outcomes, which could yield lower health and life insurance premiums."[23]

Another argument against federal legislation proposed by Thomas Boyden of the Ayn Rand Institute is that insured health coverage is not a right, rather a value.[24] Furthermore, Llewellyn H. Rockwell, Jr. of the Ludwig von Mises Institute argues an advantage of DNA sequencing is that it allows insurance companies and businesses to discriminate between individuals rather than spread the cost of a disease across all insurance holders.[25] As a result of insurance discrimination laws, everyone's insurance premiums increase, which violates others' rights.

When the insurer is prohibited from using certain categories of information to set prices, it results in regulatory adverse selection or anti-selection. Given that risk factors increase health care costs, the insurer has a reduced ability to accurately manage risks. An individual's demand for insurance correlates with the individuals risk or loss, consequently, since bad products or bad customers are more likely selected.

To counter the effects of adverse selection, insurers, to the extent that the law permits, may request additional medical information and raise pricing based on risks. Insurers can also provide favorable rates for those with healthy lifestyle choices such as fitness center memebership or abstaining from smoking.

A DOUBLE STANDARD

In an attempt to define genetic property rights, federal regulators have passed legislation to provide safeguards for the use of DNA samples. Educational programs which are part of ELSI inform genomics researchers of the implications and consequences of their work. GINA protects individuals providing DNA samples protection from individual health insurance or employment discrimination. However, the government that makes the rules does not necessarily follow them.

For example, the FDA has an exemption for informed consent during the clinical trials of blood substitutes. Following the Vietnam War, researchers tested blood substitutes for use on soldiers on battlefields to prevent shock resulting from low blood pressure due to large volumes of blood loss. Because blood requires typing and matching to avoid fatal clotting, blood transfusions are not readily available on the battlefield or at accident sites. Also, blood has a forty-two day shelf life and requires refrigeration.

The standard procedure is to use saline solution to restore blood volume and increase the likelihood of survival after traumatic injury. If organs do not receive enough oxygen, organ damage or death usually occurs. Although blood substitutes do not have white blood cells found in whole blood, they have hemoglobin to carry oxygen derived from animals and expired human blood.

The FDA has placed a number of blood substitutes in clinical trials. Northfield Laboratories developed PolyHeme to overcome the drawbacks of working with whole blood. It has a shelf life of over twelve months while stored at room temperature, is synthetic but purified to minimize risk of viral disease transmission, and is compatible with all blood types.

To test experimental blood substitutes, selected hospitals use trauma victims in life threatening situations for clinical trials, which presents a challenge to obtaining informed consent since patients are in trauma or are unconscious. In 2006, those living in 32 communities in 18 states, and anyone traveling through these communities were potential guinea pigs without consent in a Phase III study. To avoid participating in the clinical trials, citizens had to inform their local testing site and wear a medical bracelet. PolyHeme is the fifteenth experiment allowed by the FDA for emergency medical trials exempted from informed consent.

In the PolyHeme clinical trials, in circumstances where a patient at least eighteen years old is in critical condition at the scene of the injury, random patients either received saline solution or PolyHeme for up to twelve hours as needed after transported to selected trauma centers. Results from the clinical trials revealed more heart attacks and deaths in the PolyHeme patients when compared to a control group. Northfield Laboratories was able to withhold proprietary data relating to the trials. The data went unpublished because the labs had no government requirement to report problems.[26]

Unfortunately, similar to Polyheme, the other potential blood substitutes have either failed or were later withdrawn due to side effects including liver and kidney failure, irritated blood vessels, and heart attacks. One alternative to avoid human risks is to discontinue human testing and return to lab animals.

The federal government provides exemptions to federal agencies for individual health insurance or employment discrimination. For example, the U.S. military and Veterans Administration are exempt from GINA. The armed forces specifically deny disability benefits to service people with congenital and heredity conditions unless they have eight years of service.[27] The rationale for this exclusion is to protect the armed forces from becoming a magnet for people who know they will have costly genetic diseases.

Another example where genetic testing may have practical applications is the space program, specifically for astronaut screening. In the Bioastronautics Roadmap report, the National Academy of Sciences (NAS) critiqued NASA's failure to deal with issues of human sexuality.[28] This was a factor in NASA shuttle astronaut Lisa Nowak's alleged breakdown. Nowak received notoriety when she attempted to kidnap a rival in a love triangle while wearing an adult diaper.

NASA's current screening process for psychological traits involves a battery of tests and extensive interviewing. Long space flights elevate the stress hormone cortisol which can trigger mood swings or depression. The NAS report recommends stress-response tests in the astronaut selection process. Since some behaviors have a genetic component, NASA may consider genetic screening in the selection process.

Genetic screening of NASA astronauts is also useful for long-duration space missions where it is not practical to dispose of dead bodies. In a study on 295 astronauts, scientists determined there is a higher rate of cataracts and cancer resulting from space radiation.[29] NASA is considering genetic screening on potential astronauts for susceptibility to space radiation induced-cancers.

CHAPTER 3

Out of the Woodwork

For the public to receive benefits from biotechnology and biomedical research, the legal infrastructure is currently set up so that researchers invent useful commercial products, acquire a patent, and create profits for company stockholders. In theory everybody wins: the public, investors, and inventors. However, everyone does not perceive all biotechnology patents as a win-win situation. Consequently, the current legal infrastructure has become a divisive public policy issue.

Over the past several decades, activists have come out of the woodwork in protest. Activism has led to a movement advocating patent reform or eliminating the commercial model in favor of a commons model for biotechnology and biomedical research.

If biotechnology patents are intrinsically bad, why weren't activists viscerally opposed to the earlier inventions that led to intellectual property protection? In 1873, the French government granted Louis Pasteur patents on a purified yeast cell and for a process for making beer. In 1941, the U.S. government patented penicillin which is a naturally produced antibiotic. In 1873, the French government granted Pasteur patents on a purified yeast cell and for a process for making beer. In 1941, the United States government patented penicillin, which is a naturally produced antibiotic.

To understand the current biotechnology infrastructure, it is necessary to understand the historical events that have lead up to it. Among the historical factors that have led to the current biotechnology infrastructure are the need for incentives to inventors, a shift in norms from basic to applied research, the emergence of venture capital, legislation leading to a mandate for the transfer of technology to the private sector, and favorable rulings by the courts and United States Patent and Trademark Office

(USPTO).

Activists criticize the current legal infrastructure in biotechnology from two directions: economic and moral. Based on these historical factors, my analysis of the economic criticism that taxpayer funded inventions should remain public property is actually the result of government failures that require further government intervention. Several studies reveal that a second economic argument, the tragedy of the anticommons, does not have evidence to support its claim.[1] My analysis of the moral criticisms that DNA is a discovery, not an invention, that patenting living things is immoral, and that "ownership" of genes is immoral reveals that these claims are based on misperceptions related to patent law.[2]

A MANDATE FOR PRIVATE TRANSFER

Although known for his vaccines for anthrax, rabies, and chicken cholera, Pasteur was unable to obtain patents on them. This is because an 1844 French law prevented the patenting of pharmaceutical preparations and remedies.[3] Without patents for his vaccines, Pasteur had a monopoly indefinitely and used the profits to keep the Pasteur Institute financially independent. The Pasteur Institute was under no obligation to reveal its vaccine to the public after a stated period of time. Because of the time and costs to develop the vaccines, he kept his inventive processes as trade secrets. As a result, this required other biologists to learn directly from the Pasteur Institute or to develop the process of vaccine production themselves. However, a requirement for a patent is a written description that enables skilled persons to practice the invention assuring the public receives the benefits in exchange for the temporary exclusive rights.

Without clearly defined property rights, another scenario that can occur is medical applications are not pursued because of the lack of incentives. In 1928, Scottish bacteriologist Alexander Fleming accidentally discovered penicillin after leaving a Petri dish of *Staphylococcus* bacterial culture uncovered for several days. He later found it contaminated with mold. Fleming then discovered the mold was dissolving all the bacteria surrounding it.

In 1935 Oxford University researchers Howard Florey and Ernst Chain obtained a culture of Fleming's original mold and were able to separate and purify it. After testing penicillin's antibiotic properties on animals, they found it was non-toxic, unlike quinine and arsenic. Because Fleming did not file for a patent, private companies did not have an incentive to pursue drug development. As a result, the antibacterial properties produced by molds in penicillin did not become commercially available until 1941.

The U.S. government sponsored research for mass production of penicillin using fermentation, and subsequently it was used to treat infections on wounded soldiers during World War II. In 1948 the U. S. government licensed the patent on penicillin to several firms that would mass produce the drug and compete in the market which subsequently drove its price down.[4] Dorothy Crowfoot Hodgkin used X-ray crystallography to determine penicillin's chemical structure, which later enabled scientists to synthesize penicillin in the lab. It is now routinely used as an antibacterial treatment for surgery and organ transplants, and for the treatment of burns, syphilis, meningitis, and pneumonia.

In contrast to the situation with penicillin, at the end of the 1970s, federal agencies held roughly 28,000 patents; however, they had licensed fewer than 5 percent of them.[5] Consequently, citizens were not receiving benefits of federal research a government failure. Since part of the public sector's role is to reduce barriers that cause private companies to fail, the government passed legislation mandating the private transfer of biotechnology research to provide of a more efficient infrastructure. Congress intervened by passing the Bayh-Dole Act (1980) requiring universities, non-profits, and small businesses to patent and commercialize federally-funded research and inventions.

Technical knowledge and capital are necessary to ensure the success of biotech companies. In response, Congress also passed The Federal Transfer Technology Act of 1986 to amend the Stevenson-Wydler Technology Innovation Act of 1980 which requires federal government laboratories to seek opportunities to transfer technology to industry, universities, and state and local governments through technical assistance, grants, and collaborations. In addition, Congress revised the Federal Tax Code to encourage investment in private and university research through limited partnerships.

A SHIFT IN NORMS

In *Pasteur's Quadrant*, Donald Stokes points out how the emphasis of scientific research has changed historically. Pasteur's microbiology research on vaccines, fermentation, and pasteurization was both basic and applied science. Historically, university research followed the Bohr model of basic science research named from research on the atom. After the Cold War era, science and engineering followed what Stokes refers to as Pasteur's quadrant, with both basic and applied science. Today, most industrial research has an emphasis on commercial inventions, which falls into Edison's quadrant. The shift to a greater emphasis on applied science has resulted in increased patent protection.

Table 3.1 Comparative models of emphasis in scientific research[6]

Bohr's Quadrant basic science	**Pasteur's Quadrant** basic and applied science
Peterson's Quadrant Peterson's field guide of bird markings for birdwatchers is neither basic scientific research nor applied science.	**Edison's Quadrant** applied science

Billions of dollars of private and public funding is allocated annually for medical research. An argument that activists use in favor for the commons model is that taxpayer funded inventions belong to the public. However, government funding of basic research only covers part of the cost of developing drugs, therapeutics, and diagnostic tools. In the process of finding new drugs, and developing diagnostic tests and vaccines to treat human diseases, biotech companies rely on patents for royalties for research and development costs.

For drugs, the cost for the twelve to fifteen years of discovery and approval through the clinical trials process is mind-boggling. Steven Burrill's 2003 edition of Biotech Life Sciences: Revaluation and Restructuring: 17th Annual Report on the Industry provides the cost breakdown for drug development:[7]

- discovery: $30,000,000
- pre-clinical: $30,000,000
- phase I and II clinical trials: $50-100 million
- phase III clinical trials: more than $100 million

In 2003, the average cost needed for the development of a blockbuster drug — one which has over $1 billion dollars in sales — was over $800 million.[8] The costs do not stop with the launch of a drug; manufacturers are required to follow the results of significant adverse experiences. Significant capital is required from investors for research and development to invent and manufacture drugs, and to proceed through the regulatory approval processes. Biotechnology companies will not take the risk of investing in drugs or diagnostic tools that can benefit patients unless they can derive profits from the successful drugs.

Scientists have an incentive not to share with others because of the free-rider problem. Other parties can benefit from an invention although they did not share the costs of product development. According to Michael Mireles, "Patent law is designed to correct a market failure wherein too few inventions are created because copyists may easily free-ride on the efforts of inventions."[9]

In a free-market economy, patents encourage inventions by providing economic incentives to inventors through encouraging capitalists to invest in commercialization while excluding others. Although activists complain that patents limit what other parties can do, that is exactly what the framers of the Constitution intended patents to do. Patents also ensure the inventions are made available to the public rather than encouraging inventors to employ trade secrets or to not pursue inventions.

WHAT IS PATENTABLE?

Our Founding Fathers promoted patents to provide economic benefits to inventors and to provide products that increase the quality of our lives. In 1793 Thomas Jefferson wrote a statute, Article I, Section 8 of the U.S. Constitution, giving general guidelines of what is patentable:

> Congress has the power to promote the progress of science
> and useful arts by securing for a limited time to authors and
> inventors the exclusive right to their respective writings and
> discoveries. Patents can be obtained for any new or useful
> art (replaced with process in 1952), machine, manufacture,
> or composition of matter, or any new or useful improvement
> thereof.

When the Founding Fathers wrote the Constitution, clearly biotechnology patents were not specifically addressed. Since the Constitution is silent on this topic, public policy issues are left to the three branches of government to resolve. The Supreme Court has established three categories which are not patentable: laws of nature, natural phenomena, and abstract ideas. Therefore, these categories belong in the public domain. A series of legislative, judicial, and USPTO rulings have determined molecules, plants, stem cells, and animals are patentable.

MOLECULES

Almost a century ago the courts settled whether or not a chemical compound is patentable. Parke-Davis developed and patented a pure form of adrenaline called epinephrine. A rival company challenged the patent in *Parke-Davis v. H.K. Mulford* (1911) on the grounds that a natural product is not patentable. The ruling in favor of Parke-Davis stated that isolated and purified forms are different from the naturally occurring chemical state and set the precedent for U.S. patent law. Subsequently, the federal government's penicillin, Genentech's insulin, Parke-Davis's epinephrine, Merck's vitamin B12, and Amgen's EPO have all received patent protection.

Patents also provide an incentive for venture capitalists to invest in commercialization and to fund biotechnology start-ups because of their potential for high rates of return. The Cohen-Boyer process patent for genetic engineered recombinant DNA is a classic model of technology transfer described in the Bayh-Dole Act. By isolating the gene coding for these naturally-occurring molecules and transferring them into bacteria, pharmaceutical companies can produce drugs in large quantities. This patent led to a number of blockbuster drugs including synthetic insulin, EPO, interferon, and human growth hormone.

In 1976, Silicon Valley venture capitalist Robert Swanson with Kleiner & Perkins approached Herbert Boyer at University of California Medical School who had discovered and patented restriction enzymes with a business proposition to manufacture drugs. With $500 each, they agreed to start Genentech which would market and sell Cohen and Boyer's genetically engineered products. Swanson raised additional money for staffing and lab equipment.

Using Boyer's work, which was basic science, investors were convinced genetic engineering a synthesized form of human insulin would provide a less expensive method. In 1982 Genentech created a synthetic form of insulin called Humulin which became the first biotechnology drug approved by the U.S. Food and Drug Administration. Prior to genetically engineering, drug companies obtained insulin through the expensive extraction process from the pancreas of cattle, pigs, sheep, and human cadavers.

When Amgen created EPOGEN, they paid royalties to Cohen and Boyer for their genetic engineering patent and received a patent on the drug itself. The genetic engineering patent issued in 1980 generated over $190 million from licensing revenues for Stanford and the University of California at San Francisco from $35 billion in worldwide sales, before the patent expired in 1997.[10] One-third of the revenues went to the investigators, Cohen and Boyer, who chose to use the money for further research.

Over four-hundred firms have licensed the rights to use the genetic engineering patent. In 1979 Genentech cloned the gene for human growth hormone for children with growth deficiencies. In 1980, Biogen cloned interferon, a chemical normally produced by human cells in response to viral attacks. This drug promotes the production of a protein that stimulates the immune system. Genetic engineering enabled scientists to provide an inexpensive and abundant supply which is used for the common cold, hepatitis, to combat viral infections in transplant patients, as well as an anticancer drug.

In 1983, Amgen cloned erythropoietin in the form of recombinant erythropoietin (EPO) for anemia and dialysis patients. In 1987, Amgen secured a drug patent under the name EPOGEN. Since 1989, when the drug was approved, through 2007 Amgen has sold over $25 billion worth of EPOGEN.[11] Amgen received seven US patents based on the work of scientist Fu-Kuen Lin, including the EPO gene, genetically engineered cells for making EPO, processes for manufacturing, the drug, and for methods of treatment.

To many people, the idea that DNA sequences are inventions, rather than discoveries is counterintuitive. In *The Common Thread*, John Sulston raised objections to gene patents on moral grounds, since humans have a common genetic heritage.[12] However, in patent law the interpretation of the Framers' use of discovery refers to inventions that are discoveries requiring human innovation.

Claude Shannon and Warren Weaver, authors of *A Mathematical Theory of Communication*, made an important distinction regarding the computational communication of information, recognizing the difference between electronic signals and thoughts. These properties also are relevant to DNA. The dual nature of DNA consists of a chemical compound and a template for the transfer of information. DNA patents are on the molecule, not on the genetic code or the instructions encoded in the sequences of nucleotides.[13]

In the early 1990s, researchers at NIH filed for patents on thousands of partial sequences of genes are typically 300-500 base pairs called expressed sequence tags (ESTs). Reid Adler, the head of NIH's Transfer Technology Office approached Craig Venter and told him that NIH was obligated under the Bayh-Dole Act to try to patent his ESTs.[14] The USPTO decides which patent applications to approve using the criteria of novelty, utility (usefulness), and non-obvious to those skilled in the art. The EST sequences research utility is that of a probe in order to find the full length gene. ESTs are also used to isolate promoters, polymorphisms, and as markers on genetic maps. In 1994, the NIH agreed to retract its patent applications. However, in 1998 the USPTO issued the first controversial EST patent to Incyte Pharmaceuticals on a human kinase. In 2001, the USPTO revised the utility guidelines requiring a potential patent's utility to be "significant and substantial." So, the use as a probe no longer satisfies the USPTO's utility criteria.

In the 1980s the search for a genetic link to breast cancer led to the discovery of two susceptibility genes. The mutated alleles, expressed version of the gene with a beneficial mutation, are found in statistically higher concentrations in certain families and ethic groups and do not represent the wild type or common form of the genes. Women with the mutated alleles — *BRCA1* on chromosome 17 and *BRCA2* on chromosome 13, which are DNA repair genes — have a significantly higher risk of inherited breast and ovarian cancer. Myriad Genetics patented both mutated alleles. Scientists estimate only 10 percent of breast cancers are genetic, with less than half of these related to *BRCA* mutations.

In the 1980s, in an attempt to forbid the patenting of the *BRCA* genes, activists Jeremy Rifkin and Bella Abzug led campaigns on the grounds that a DNA patent meant ownership of part a person's body. However, gene patents are not a possessory right, rather a right to exclude others from royalties.[15] Patent holders own a patent, not DNA. Myriad's patents on *BRCA1* and *BRCA2* genes simply exclude commercial rights to other parties without licensing. To own DNA would violate the Thirteenth Amendment of the U.S. Constitution adopted in 1865 which also prohibits slavery.

In 2007, Representatives Becerra and Weldon introduced a bill, The Genomic and Accessibility Act, that if enacted would ban patenting genes. The bill passed in the US House of Representatives, but did not receive enough votes in the Senate. The official press release on Representative Becerra's website for the Becerra-Weldon Bill introduced to ban gene patenting is misleading since it claims 20 percent of human genes are owned by someone.

In 2009 the ACLU filed a lawsuit in the United States Federal District Court in New York in behalf of numerous plaintiffs against Myriad Genetics and the USPTO. In the lawsuit, Christopher Hansen, a senior counsel, said the problem is with the patent office, not Myriad.[16] The ACLU claims that Myriad's *BRCA* gene patents are unconstitutional. Myriad Genetics also invented and patented a screening test. The lawsuit also claims the $3,000 fee for the screening test is financially restrictive.

Although the test is covered by health insurance, the patent provides Myriad with a temporary monopoly. The test is not licensed to other labs. Swabs and blood tests taken at labs around the world are sent exclusively to Myriad's Utah based labs for testing.

Under the current infrastructure, the government can require a reasonable pricing clause. Under the Bayh-Dole Act, a federal agency has the right to march-in and force a recipient of a federal grant to non-exclusively license patented technology to meet society's needs. While Myriad, which shares the BRCA gene patents with the University of Utah, did not receive government funding, one of the company's founders, Mark Skolnick, initially worked with NIH and the University of Utah on isolating the genes.[17] The government has chosen not to exercise its march-in rights.

PLANTS

Three plant patent rulings have enabled scientists to cross-breed and genetically modify agricultural and horticultural products for desired traits and receive patent protection. As a result, we enjoy human inspired flowers and foods that do not occur naturally.

In the 1920s, plant breeders sought patent protection for hybrid seed development. Congress passed the Plant Patent Act of 1930, making patents obtainable on asexually reproduced plants through processes such as budding, grafting, and tissue culture. Four decades later, Congress passed the Plant Variety Protection Act of 1970 for sexually reproduced plants. The two plant acts made a distinction between breeders and inventors.

In 1985, the USPTO Board of Patent Appeals ruled in *ex-parte Hibberd* that plant are protectable by utility patents. Utility patents can apply to the method used to engineer a plant, the genetic sequences that are inserted, and the resulting plants. An example of a utility patent is the Terminator patent which renders seeds sterile.[18] This type of patent allegedly creates hardships for small rural farmers that depend on the patented plants seeds. The Supreme Court ruled that farmers must repurchase the plants annually since the utility patent provides no exemption for farmers or plant breeders.[19]

STEM CELLS

In 1998 James Thomson, a biologist at the University of Wisconsin, was the first to isolate human embryonic stem cells; cells in embryos that can develop into specialized human cells. Geron Corp. and the Wisconsin Alumni Research Foundation (WARF) supported Thomson's research since it was ineligible for federal funding. Based on Thomson's work, the

USPTO issued three patents to Geron and WARF; the development of a cell culture of human embryonic stem cells derived from pre-implantation embryos and the techniques used for maintaining the cells.

In 2006 the Public Patent Foundation and the Foundation for Taxpayer and Consumer Rights challenged these patents on the basis the methods used for isolating a human stem cell line impede research and were obvious based on a 1990 paper and two textbooks. They argued that the patent claims did not pass the USPTO's novelty test since there was not a significant advance over what was previously known in the field. The techniques used by Thomson were the same used to make mouse embryonic stem cell lines in 1981. In 2008, the USPTO reexamined the patents held by WARF and upheld them. However, the European Patent Office rejected the human stem cell line patent.

In response to the Public Patent Foundation and the Foundation for Taxpayer and Consumer Rights' allegations that the patents inhibited stem cell research, WARF narrowed the scope of the stem cell line patent to those derived from pre-implantation embryos, excluding those derived from fetal tissue or therapeutic cloning. Carl Gulbrandsen, managing director of WARF, says he is a proponent of disseminating the technology to researchers and that restrictions in federal funding for stem cell research are far more obstructive than the WARF patents.[20] Geron does not require academic researchers to pay royalties, but does charge for licensing to commercial users.

ANIMALS

In the 1970s Ananda Chakrabarty, a microbiologist working for General Electric, developed bacteria capable of breaking down crude oil with the industrial application of cleaning up oil spills. Chakrabarty's group discovered that plasmids, the circular DNA of certain bacteria, are capable of degrading camphor and octane, two components of crude oil. In addition, the researchers developed a process to transfer and maintain stably four different plasmids (circular DNA found in bacteria) which are capable of degrading oil, to a single *Pseudomonas* strain, which itself has no capacity for degrading oil.

> Chakrabarty successively mated one strain of bacteria with another, each containing plasmids desirable for his purpose, then fused the plasmids together with X-rays to ensure that they would coexist stably in the cell. This was not genetic engineering in the sense of using the technique of recombinant

DNA. Nor was it entirely conventional mating. Key to the inventive step was the fusion of the plasmids by X-rays.[21]

Chakrabarty applied for two patents: one for a process to make oil eating bacteria and the other on the bacteria themselves. In 1980, the USPTO granted the patent claims for the process which produced the oil eating bacteria. The the product patent was rejected in 1974 because the bacteria are products of nature and are living. The Patent Office Board of Appeals agreed with the examiner. The Court of Customs and Patent Appeals overturned the ruling of the Board of Appeals in 1976, reasoning that because the bacteria were alive does not forbid patenting.

In *Diamond v. Chakrabarty* (1980), the U.S. Supreme Court upheld the Court of Customs and Patent Appeals ruling. The high court ruling stated the distinction is not between living and non-living things, but between products naturally occurring, whether living or not, and human made inventions. Citing the Patent Act of 1952, the court concluded nothing precludes the patenting of nature. Chief Justice Berger noted in the majority opinion, Congress intended "anything under the sun" that is made by man to be patentable. The product patent was issued in 1984.

In 1987, the USPTO Commissioner announced as the result of *ex-parte Allen*, higher animals are protected by utility patents. In 1988, the USPTO granted the first patent for a transgenic animal to Harvard University and DuPont for the OncoMouse. The OncoMouse has a DNA sequence making it prone to cancer. This genetically modified mouse enables researchers to better understand environmental factors that trigger cancer development. Harvard University biologists Philip Leder and Timothy Stewart developed the OncoMouse inserting a known oncogene into a preexisting strain of mice using an already known technique, the microinjection of DNA. Critics of the patent claimed that inserting a gene into a life form was already obvious.

Researchers may use the OncoMouse for a licensing fee. In addition, the licensing agreement contains a reach-through provision that requires royalty payments on sales of products that were developed from subsequent downstream discoveries.[22] DuPont has offered non-commercial research licenses on the terms that DuPont has the right to develop commercial products resulting from the use of the licensed mouse.

A second area of contention with the OncoMouse is that the USPTO issued a broad patent which includes the patents on similar inventions which has angered activists and researchers. Although a mouse and one particular activated oncogene were used to develop the OncoMouse, its patent applies to any non-human mammal containing an activated on-cogene. Consequently, if researchers create similar animal models with an oncogene, they would have to pay royalties to Harvard and Dupont on profits from sales and any subsequent medical breakthroughs derived from it. The USPTO no longer grants patents that are as broad as the Onc-oMouse patent.

THE TRAGEDY OF THE ANTICOMMONS

While some activists are advocating patent reform, others are challenging the legal infrastructure in support for a biological commons. The United States Bill of Rights clearly outlines our individual rights; however, it has no references to the collective good. During the Human Genome Project, researchers receiving government funding took it upon themselves to de-posit genomic data in public databases such as GenBank under voluntary guidelines referred to as the Bermuda Rules. Some researchers have devel-oped consortiums voluntarily placing data such as SNPs and haplotypes in open-source databases.

Critics of the commercial model have taken this action because they believe patents hinder progress due to prohibitive fees and a backlog of patent applications slows the process of discovering cures for diseases. Mi-chael Heller of Columbia Law School and Rebecca Eisenberg of the Uni-versity of Michigan Law School argue that competing patents in biomedi-cal research can also prevent the creation of useful products. Heller and Rebecca Eisenberg refer to multiple property claims to a single gene or licensing royalties as the Tragedy of the Anticommons because a resource may become underutilized.[23]

According to Ted Buckley, the former Director of Economic Policy with the Biotech Industry Association, the Tragedy of the Anticommons is theoretical and lacks empirical evidence. If a tragedy of the anti-com-mons took place, it would result would in a decrease in scientific research; however, the exact opposite has occurred. Research & development in aca-demia and biotech companies has increased.[24] Since 1988 R&D of publicly traded companies has increased 60 percent, from 1995 to 2005 venture capital funding for biotechnology companies increased 300 percent, drug applications to FDA have increased, from 1998 to 2005 biological com-pounds entering preclinical trials increased 37 percent, and no academics have reported abandoning research due to patents.[25]

The NIH asked the National Academy of Sciences (NAS) to study the effects of patenting DNA on research and innovation. In two studies, the NAS also disputed Heller and Eisenberg's Anticommons theory.

According to a 2003 study:

> Changes in technology and policy over the last several
> decades have led to concerns that the patent system may be
> creating difficulties for those trying to do research in bio-
> medical fields. However, we find that drug discovery has not
> been substantially impeded by these changes. We also find
> little evidence that university research has been impeded by
> concerns about patents on research tools. Rather researchers
> managed to adopt solutions or work around the problems
> through licensing, going offshore, using public data, legal
> action, and infringement.[26]

According to a 2005 study:

> A committee found that the number of projects abandoned
> or delayed as a result of difficulties in technology access is
> reported to be small, as is the number of occasions in which
> investigators revise their protocols to avoid intellectual prop-
> erty issues or in which they pay high costs to obtain intellec-
> tual property. Thus, for the time being, it appears that access
> to patented inventions or information inputs into biomedical
> research rarely imposes a significant burden for biomedi-
> cal researchers. In fact, only 1 percent of the 414 biomedical
> researchers interviewed reported suffering delays because of
> patents.[27]

Part of the reason for this finding is that the federal government has
proactively intervened to eliminate barriers for academic researchers and
in specific situations for private companies. For academic researchers the
major concern regarding intellectual property is for experimental use and
for pharmaceutical companies a major concern is research and development prior to the expiration of patented drugs. Several examples of litigation and amendments to regulations demonstrate that striking a balance
between protecting inventor's rights and providing an infrastructure that
also allows innovation is a dynamic process.

In *Roche v. Bolar* (1984), when Bolar Pharmaceutical, a generic man-
ufacturer, used Roche's Valium to research a bioequivalent drug, Roche
claimed Bolar had infringed on its patented drug, but the Federal Court
disagreed. The Court of Appeals rejected Bolar's argument that it used the

patented drug compound to perform bioequivalency tests to obtain Food and Drug Administration (FDA) approval since their drug research was for business purposes. So, the decision extended Roche's monopoly beyond its expiration date. The Appeals Court acknowledged that the decision would create an extension of Roche's patent term, but concluded that this was Congress' intention because of the wording in the Patent Act and also the Food, Drug, and Cosmetic Act.[28]

Congress responded to the Appeals Court's ruling in *Roche v. Bolar* with the Hatch-Waxman Act (1984) which reversed the decision. An amendment in the Patent Act allows generic drug companies to experiment with patented brand name drugs to develop generic drug substitutes and obtain FDA approval of the generic drugs prior to the patent expirations of the brand name drugs without infringement. The origin of the common law experimental use exemption to patent infringement is attributed to Justice Story's opinion in *Whittemore v. Cutter* (1813).[29] In this case, Justice Story stated that "It could never have been the intention of the legislature to punish a man who constructed a patented machine merely for philosophical experiments."

The Hatch-Waxman Act provides a narrow experimental use exemption or safe harbor from patent infringement. However, Congress did not intend for the experimental use exemption to violate the patent holder's rights. Consequently, in some cases exemptions from patent infringement are denied.

In *Madey v. Duke University* (2002), a Federal Appeals Court ruled that after firing Madey Duke's use of his patented free electron laser oscillator furthered the universities business objectives. Duke's research failed the experimental use exemption since it served the purpose of educating students and faculty participating in the research projects, enhancing the status of the university, and luring lucrative research grants, students, and faculty to the university.[30] The broad interpretation of the Federal District Court experimental use analysis was now narrowed not by non-profit status but rather by legitimate business objectives.

In another landmark patent infringement suit *Merck v. Integra* (2005), a jury awarded Integra $15,000,000 in royalty damages. Integra had patented several peptide sequences used for healing wounds, but was unable to develop commercial products. Merck funded research using the patented sequences that entered clinical trials. The Federal Appeals Court upheld the ruling since Merck's research was also used to identify new pharmaceutical compounds. According to the Appeals Court, the Hatch-Waxman Act experimental use exemption was to reverse the Roche v. Bolar decision and not to deprive patent protection to inventions.[31]

Under the current infrastructure, defining property rights provides an efficient economic system by providing incentives. Western Europe, Japan, and the United States which have the world's major patent offices also have high standards of living and are the leaders in invention and production of medical and high tech products, partially due to their intellectual property infrastructure.[32]

Scott Kieff, a Patent Law Professor at Washington University who studies institutional economics (the economic influence of institutions, laws, and norms), points out industrialized nations have accumulated wealth through intellectual property rather than from resources.[33] According to Kieff, this is how Japan, a small island with limited resources, became an economic superpower.

If patent reform is to occur, it should promote innovation through a more efficient regulatory structure, not deter it. It is not clear how taking away incentives from inventors will lead to medical cures.

CHAPTER 4

The Rise & Descent of Vitalism

Roughly 50,000 years ago a number of our ancestors migrated out of Africa, and our species proceeded to inhabit almost every portion of the earth. At some point these distant relatives were curious about mankind's relationship to the universe. They sought explanations for the origins of life, human origins, inheritance and the variation of traits, speciation, diseases, and the formation of the earth. For most of modern human existence, from the ancient civilizations through the enlightenment, religious dogma provided the explanations to these puzzles.

Early civilizations, including the Sumerian, Greek, Roman, Indian, and Norse believed powers were in the hands of supernatural deities. American Indians and Polynesians practiced totemism, a belief that objects have powers to provide protection from evil gods and spirits. A number of cultures including the Chinese practiced animism, a belief that animals possess supernatural powers.

As the earliest civilizations developed, so did traditional forms of medicine. In a number of early cultures, healers were men with inherited medicinal powers referred to as Shamans. Chinese traditional medicine and philosophy is based on a force or energy called *chi* (Mandarin) and *qi* (Cantonese) moving through channels in our bodies called meridians. Chinese healers treat diseases by determining the imbalance of forces, referred to as *yin* or *yang*, in the meridians. To restore a balance in the meridians, healers prescribe the consumption of foods that contain either positive or negative energy. Similarly, in traditional Indian healing referred to as Ayurvedic medicine the vital forces are called *prana*. These traditional forms of medicine are the basis for modern naturopathic and holistic medicines which use acupuncture and herbs to balance meridians.

Historians and philosophers credit Aristotle with creating a branch of philosophy known as natural history in the fourth century BCE. Aristotle based his explanations to the puzzles of the universe on vitalism. Vitalists believe the processes of life are not explained by the laws of physics and chemistry, rather life is partially explained by a self-determining nonmaterial force.

Religions have creation stories to explain how the earth and its life forms came into existence. In the case of Christianity, God created the universe roughly 6,000 years ago. As stated the first book in the Old Testament in Genesis 1:1-2, this took place over a seven day period. On the second day God created the firmament, on the third day God created plants, on the fifth day God created birds and sea creatures, and on the sixth day God created land animals and man in his image. In an act of spontaneous generation, Adam was created out of dust, and using Adam's rib, the female of the human species emerged.

During the 17th century Age of Reason and the 18th century Enlightenment, scholars began questioning traditional doctrines and called for human progress using the empirical method including science and reason. In this period, research by natural historians provided explanations to the universe and life forms that questioned vitalism (see Table 4.1). These findings lead to the descent of vitalism.

Table 4.1 Scholars in the 18th and 19th centuries making discoveries that contradict vitalism and creationism

Copernicus and Galileo: the solar system
Cuvier and Lyell: paleontology, geology, and taxonomy
Pasteur: plagues
Darwin and Mendel: differentiation in life forms

THE SOLAR SYSTEM

In Aristotle's time the understanding of the solar system was quite different than it is today. The prevailing theory formulated by the Greek astronomer Ptolemy in AD 150 stated the earth is stationary and the sun revolved around it. The books of Psalms 93 and 104, and Ecclesiastes 1: 5 mention the motion of the celestial bodies and the suspended position of the earth.

In 1543, astronomer Nicholas Copernicus formulated the heliocentric model of the universe. Copernicus could not provide evidence for his model; however, his writings were pivotal to the Scientific Revolution. In

1610, based on observations with a telescope, astronomer Galileo Galilei was able to provide evidence for the heliocentric theory. Using a telescope invented by a Dutch spectacle maker, Galileo made crucial observations that proved the Copernican hypothesis. He observed the Sun had dark patches now referred to as sunspots. He further observed motion of the sunspots indicating the Sun was rotating on an axis. His observation contrasted the doctrine of perfect heavenly bodies and implied that the earth may also rotate.

At the time, science was not independent of politics and the church. The religious community did not welcome this alternative understanding of the universe. Because, if the Aristotelian view of our solar system was wrong the Roman Catholic Church was wrong. Challenging the church's view was heresy. The Catholic Church placed Galileo on trial for teaching, defending, and discussing the Copernican heliocentric theory which contradicted church dogma.

After Galileo's house arrest he argued in a letter in 1616 that if both scripture and nature are divine creations, it is then impossible for an interpretation of scripture to be correct if what we observe in nature is contrary. Pope Urban VIII's Jesuit advisors recommended a judgment of heresy from the royal court. Following a second trial in 1633, this time after a hearing by inquisition, the Catholic Church censured Galileo, gave him a lifetime house arrest, and placed him under surveillance. It was not until 1983 that the Catholic Church formally acknowledged Galileo was right.

In 1788, Buffon revised the geologic time scale placing the world much older than the 6,000 years claimed by the church. This was important because it meant species could change and were not created perfectly by God. Based on the time it would take the earth to cool, Buffon estimated the earth was roughly 75,000 years old. In 1862, Lord Kelvin revised the timeline to over 20 million years old based on cooling.

PLAGUES

As civilizations arose and population density increased, our ancestors sought explanations for the mysterious diseases that decimated populations. These illnesses referred to as plagues have symptoms similar to influenza including fever, chills, and swollen lymph nodes, but are fatal. In the sixth century the Justinian Plague claimed millions of lives in the Byzantine Empire. The Black Death swept through Europe in 1348 killing a third of the population.

During the time of the ancient Roman and Greek civilizations, the idea that living things could arise from non-living things originated. In the fourth century BCE, Aristotle believed animals could originate from the animating force *pneuma*, rather than from eggs or copulation, a phenomenon referred to as spontaneous generation. It was a commonly held belief among natural historians that maggots could spring to life from decaying meat and sweat. Vitalists believe invisible particles dispersed through the atmosphere provided the mechanism of delivery. Malaria is derived from Italian (*mala aria*) meaning bad air.

In the first millennium or the Middle Ages a likely explanation for infectious diseases was punishment from the Gods for past sins committed. Book I Samuel 5:6 of the Hebrew Bible describes how the Philistines were struck with a plague for stealing the Ark from Israel. This passage provides the basis for evangelist Pat Robertson's belief that Hurricane Katrina (2005) and the Haiti earthquake (2010) were sent by God as punishment for the decadence that took place in New Orleans and Haiti.

In 1668, Francesco Redi an Italian physician and naturalist understood that maggots developed from eggs laid by flies. Using a controlled experiment, he tested his belief. He placed meat in a covered flask, it decayed and became putrefied, but no maggots appeared. However, when he left the flask open maggots appeared, suggesting that something traveling in the air was responsible for the maggots. In spite of the finding, the belief in spontaneous generation remained popular.

In 1859, the French chemist Louis Pasteur experimented with microorganisms to determine if they were the source of infection. He placed a boiled nutrient broth in a flask with a neck to trap water, preventing contact with the atmosphere. Similar to Redi's observation, no maggots appeared. However, when using a flask with a straight neck exposed to bacteria and fungi, they did appear. This implied that plagues were not caused by spontaneous generation or by a divine source, rather by microorganisms.

Through similar experimentation roughly two centuries later, Pasteur's germ theory became the prevailing belief among scientists and subsequently led to the branch of science called microbiology. In 1894, based on the germ theory, Swiss bacteriologist Alexandre Yersin of the Pasteur Institute discovered the bacteria *Yersina pestis* caused the Bubonic plague which killed millions of people. The Bubonic plague is fatal because the infection caused by bacteria introduced by flea bites spreads either to the host's lungs or enters the bloodstream. In 1898, French scientist Paul Louis Simond established that the plague is spread by fleas which live on rodents, notably the groundhog and the black rat. In winter, occurrences of the plague disappear because fleas are dormant, however plagues may return in the spring when the fleas are active.

The discovery of vaccines prevented many deaths from diseases such as cholera, diphtheria, scarlet fever, syphilis, and smallpox also linked to microbes. Today, the World Health Organization reports roughly 3,000 cases of plagues annually with the majority of cases occurring in the Congo and Madagascar.

Pasteur's discovery has numerous practical applications. It explains why milk sours, why beer and wine ferment, and has led to the pasteurization process to destroy germs in food products and the sterilization of medical instruments. In 1862 France a crisis in the silkworm industry emerged; diseased silkworm eggs were not hatching. The French government asked Pasteur to investigate the mystery. Using a microscope, Pasteur observed the diseased caterpillars and eggs contained parasites which cause pebrine, a contagious disease which is heritable. He discovered that silk worms suffer from infectious diseases caused by protozoans, fungi, viruses, and bacteria. He was able to save the silk industry by providing healthy living conditions for the silkworms.

BIODIVERSITY

In the 18th century, natural historians began accumulating a growing body of knowledge that contradicted creation stories. In 1796, Georges Cuvier proposed that million year-old fossils of extinct species were the result of catastrophes. In 1830 Charles Lyell observed that humans do not appear in these ancient fossil records, suggesting a new timeline for the formation of the earth and the appearance of humans. In 1749 Comte de Buffon wrote 36 volumes on natural history, which discussed the similarity between humans and apes, and how regions and climates resulted in variations in plants and animals. In 1735 Carl Linnaeus wrote *System Naturae* which outlined a hierarchical taxonomic system including kingdoms divided into class, order, genus, and species based on lineage.

In 1809, Jean Baptiste Pierre Lamarck proposed two laws to explain a link between humans and other species. His first law, the use and disuse theory, states organisms adapt to the environment through modifications made through the restructuring of vital forces. According to this theory, the giraffe's long neck used for grazing leaves high in the trees became larger and stronger because it was used more frequently. For those body parts not used frequently, they deteriorated. His second law, the inheritance of acquired characteristics, states that these acquired traits pass into the next generation.[1]

In 1892 cytologist August Weismann rebutted Lamarckian inheritance arguing that the transmission of acquired characters is impossible, because the germ plasm is derived from that which preceded it.[2] According to Weismann, somatic cells are influenced by the environment; however, the germ plasm is segregated early in development and is not susceptible to environmental influences.

Weismann amputated the tails from numerous generations of mice and observed that the resulting offspring were born with normal tails. Similarly, generations of circumcised men's descendants are born with foreskins. At the time, the Weismann barrier prevailed, although the method used to arrive at that conclusion is flawed. Consequently, for a century the second law of Lamarckian inheritance, the inheritance of acquired characteristics in humans, was forgotten.

In the late 1700s, William Charles Wells, an English medical doctor, moved to Charleston, South Carolina, where he became a printer and bookseller. Following a peace treaty with Great Britain, Wells returned to England in 1784 to practice medicine. Wells later became interested in evolution. In 1813, he read a paper on the principle of evolution by natural selection before the Royal Society in London. Building on the concept of natural selection proposed by others, Wells' paper consisted of two essays which were published in 1818. In his essay titled, *An Account of a Female of the White Race of Mankind*, he proposed that those from hot climates and with darker skin are better able to resist attacks of the diseases in hot climates than those with lighter skin. Ironically, the former publisher did not aggressively market the idea.

Patrick Matthew, a fruit farmer, raised timber for naval warships by creating new varieties and eliminating those with poor quality. Matthew published *Naval Timber and Arboriculture* (1831) citing natural selection; however, he did not develop his ideas or attempt to convince others of evolutionary theory. Consequently, he also was not influential in the history of the field's development.[3]

In the 1800s the only book available to English speaking readers providing an alternative to creationism was the *Vestiges of the Natural History of Creation* (1844). Robert Chambers' book was controversial in both the religious and scientific communities. Chambers exposed readers to geological and paleontological evidence as well as comparative mammalian development which contradicted religious accounts of creation.

Alfred Wallace had a background as a surveyor, museum curator, and watch maker. Henry Walter Bates introduced the amateur naturalist to botany. They made several expeditions together collecting botanical specimens and insects, and performing fieldwork in biogeography in the

Amazon, the Malay Archipelago and the Spice Islands. In 1855 Wallace submitted a paper on evolving forms of life to the scientific journal *Annals and Magazine of Natural History*. Influenced by Thomas Malthus' *An Essay on the Principle of Populations* (1798), which discussed competition for scarce resources, Wallace proposed members of a population who are better adapted to the environment survive and pass on their traits. Wallace sent a copy of the manuscript to another naturalist.[4] That naturalist was Charles Darwin.

While training as a physician, Darwin was terrified by surgery and eventually lost interest in the profession. He transferred to divinity school and subsequently became a devout Anglican. At the time, the clergy was considered an appropriate alternative profession for his social standing. His maternal grandfather Josiah Wedgwood started Wedgwood pottery. Darwin sacrificed his professional training with yet another career change as a naturalist.

From 1831-1836, he voyaged around the world on the H.M.S. Beagle as a naturalist with Captain Robert FitzRoy. On the voyage, Darwin observed plants and animals were different from island to island, which he chronicled. On the Galapagos Islands he identified thirteen species of finches; however, only one finch species lived on the mainland of South America where it is assumed they had originated. The finch species on the Galapagos Islands differed from each other in beak shape. At first, Darwin questioned if they were separate creations or somehow related. He speculated that the various beaks were associated with competition for food sources.

While Wallace emphasized biogeographical and environmental pressures forcing adaptations, Darwin emphasized competition between individuals of the same species to survive and reproduce. Darwin used Herbert Spencer's phrase "survival of the fittest" to describe the mechanism for natural selection. The concept of fitness, that some traits provide an organism survival and reproductive advantages, implies a common ancestry of organisms.

To better understand natural selection, Darwin investigated different methods of selection. He studied artificial selection by plant an animal breeders. He wrote about sexual selection or non-random mating, proposed by his grandfather Erasmus Darwin, as a component of fitness. With a substantial investment of time and resources, females are selective when choosing mates, discriminating for or against certain traits. For example, peahens prefer peacocks with large and colorful tails which reflect the quality of their genes. In other bird species, males compete for females with songs and rituals.

Darwin formulated a mechanism for natural selection based on several existing theories. In contrast to Wallace, Darwin endorsed Lamarck's inheritance of acquired characteristics. In 1875, Darwin proposed the theory of pangenesis where particles called gemmules are shed by organs and cells, and carried in the bloodstream and collect in the germline prior to fertilization.

Darwin also endorsed the theory of blending inheritance developed by Francis Galton, meaning an offspring is the average of its parents and grandparents. In 1875, Galton proposed competition among hereditary forces later called germ cells resulted in some becoming dominant while the remaining cells rest in a dormant state which he called latent. Galton further proposed that these forces are modified only while dormant and through many successive generations.

Several years after Darwin's voyages, based on his fieldwork on biogeography, he finished a 200 page essay on evolution via natural selection as a basis for evolutionary theory. Darwin was undecided on whether or not to publish his alternative to creationism. Darwin was able to use Chambers' *Vestiges* (1844) as a lightning rod and consequently delayed the publication of his book.[5] Chambers chose not to use his name in the original publication. Chambers' identity wasn't acknowledged until the 1884 edition, which was after his death.

With support from fellow natural historians, including his friend Charles Lyell, in 1858 Darwin had his essay published as *On the Origin of Species by Natural Selection*. At the time *Origins* was published Darwin was unfamiliar with William Wells' work. However, in the fourth edition of the *Origins* printed in 1866 Darwin acknowledged that Wells distinctly recognized the principle of natural selection in his 1818 essay.

Gregor Mendel, an Augustinian monk educated at the University of Vienna with training in physics and chemistry, was also interested in the mechanism for inherited traits. In the 1860s at a monastery in Brunn, Czechoslovakia with a center for scientific studies, he used breeding experiments to investigate the correlation between genotype and phenotype.

Crossing pea plants, Mendel discovered a statistical occurrence with the inheritance a traits. In experiments with crosses of one trait, pod color, a 1:1 ratio of phenotypes resulted. Using garden peas with two traits, wrinkled and smooth pods and two seed colors, he discovered a 9:3:3:1 ratio of phenotypes resulted. This experiment demonstrated that genetic material does not blend, disproving Galton's blending inheritance theory.

While Darwin referred to particles called gemmules, Mendel was the first scientist to propose that phenotypes, or physicial characteristics, are encoded in hereditary units called factors. In contrast, Mendel found when a factor carries two different forms of a trait; one form is expressed over the other. This experiment was based on the assumption that each factor segregates independently during the formation of gametes, the reproductive cells in males and females, resulting in different combinations. Therefore, each factor has an equal probability of appearing in the next generation.

The two alleles, or forms, of each gene must segregate or separate into different cells during the formation of gametes. When gametes are produced during meiosis, reproductive cell division, allele pairs separate leaving each cell with a single allele for each trait. As a result, the gamete of each parent contains one allele of each gene. When the two alleles of a pair are different, one is dominant and the other is recessive.

In 1866, Mendel proposed three laws of inheritance creating Mendelian or classical genetics (see Table 4.2).

Table 4.2 Laws of Mendelian or classical genetics

Law of Dominance: two versions of a physical unit called factors (later named alleles), and some factors are dominant providing a statistical basis for the transmission of traits
Law of Segregation: the two factors separate into different cells so each gamete contains one factor and offspring acquire one factor (allele) from each parent
Law of Independent Assortment: each physical unit of inheritance segregates independently and has an equal probability of appearing in offspring

CHAPTER 5

How the Giraffe got Its Neck

The March 2005 issue of the *Journal of the History of Biology* was dedicated to scholarly discussions on whether or not a Darwinian revolution occurred. In order to adequately answer this question it is important to first understand how historians of science define a revolution in biology?

In Thomas Kuhn's *The Structure of Scientific Revolutions*, the physicist provides a more general analysis of revolutions in science, a model used by historians today. According to Kuhn, science texts present the inaccurate view that a series of individual discoveries and inventions have led to the present state of science. In contrast, Kuhn uses a punctuated equilibrium metaphor to distinguish between normal periods from revolutions in science (see Figure 5.1).

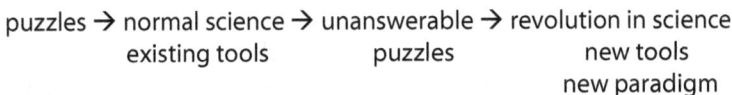

puzzles → normal science → unanswerable → revolution in science
existing tools   puzzles   new tools
new paradigm

Figure 5.1 Kuhnian Model for Revolutions in Science

Revolutions or progress in science occurs through solving puzzles over a period of which he refers to as normal science, the beliefs commonly held by scientists. These beliefs are called paradigms, which include the necessary tools. Students study these paradigms and become members of a specific profession or discipline.[1]

The search for a replacement paradigm is driven by the failure for the existing paradigm. When normal science does not explain puzzles in science, it sets the stage for a scientific revolution, a new paradigm which involves a revision in existing scientific belief. Scientists then work in a different world with a new set of tools to complement the new paradigm. Does Darwin's theory of evolution meet the standard of a Kuhnian revolution?

THE SCIENTIFIC COMMUNITY DIVIDED

Ernst Mayr argues that Darwinism is not a Kuhnian revolution for several reasons. First, no clear distinction in normal science and a revolution exists.[2] In the 1930s, Mayr astutely observed that evolutionary biology faced two major unsolved problems: adaptive changes in populations, and the origins of biodiversity or speciation.[3] Darwin not only did not have a mechanism for evolution via natural selection, he did not have a mechanism to explain how the Galapagos finches acquired unique beaks for adapting to different niches in an ecosystem and became a separate species.

In *The Non-Darwinian Revolution: Reinterpreting a Historical Myth*, Peter Bowler points out a Darwinian paradox since natural selection had little effect until the twentieth century.[4] Of Darwin's two main ideas, common ancestry and evolution via natural selection, a majority of scientists in the late 19th century accepted the first idea, but not the second. Darwin, like a number of other natural historians, had theories to explain evolution; however, they did not have a mechanism for natural selection or speciation to work, rather a series of failed theories (see Table 5.1). Thus, Darwin provided no laws and was unable to convince the scientific community.

One component of Darwin's theory of evolution via natural selection that scientists accepted is sexual selection or non-random mating. It makes perfect sense that an organism selects traits that are beneficial. Scientists also accepted that limited resources in an ecosystem will lead to competition which may not guarantee survival to those species unable to adapt.

However, the major components of Darwin's theory of evolution via natural selection were not accepted. In relation to competition for limited resources, Darwin referred to Herbert Spencer's concept of "survival of the fittest." This concept has received criticism since it is a circular argument. The fit are the ones who survive. Natural selection removes unfit variants, but fit offspring do not necessarily survive or reproduce. Fitness more accurately describes the higher probability for individuals to survive or reproduce.

Table 5.1 Components of Darwinian evolution

Accepted	Not Accepted
Common ancestry • related to apes • contradicts religious doctrine Evolution via natural selection • competition for limited resources • sexual selection or non-random mating	Evolution via natural selection • survival of the fittest • blending inheritance theory • gemmules & pangenesis • the use and disuse theory

Darwin endorsed Galton's blending inheritance theory, which Mendel later disproved. Using blood transfusions in rabbits, Darwin's cousin Francis Galton showed that gemmules, particles that carry genetic information, are not transported in blood, disproving Darwin's theory of pangenesis. Darwin also endorsed Lamarck's use and disuse theory and the inheritance of acquired characteristics in humans, which Weismann disputed. At the time, most natural historians agreed with Weismann. George Romanes coined neo-Darwinism to refer to Wallace's theory of evolution which rejected Lamarckian acquired characteristics and incorporated Weismann's germ-plasm theory.

Secondly, Mayr observed Darwinism was not a one man show. Rather, a series of minor revolutions occurred during a normal period of science, based on the attempts of Darwin's predecessors and contemporaries to better understand human origins and diversity, the mechanisms of heredity, and the relationship between organisms and the environment (see Table 5.2).[5]

The neo-Darwinians, or biometric school, led by Karl Pearson proposed that gradual and cumulative adaptations drive evolution. Without a mechanism for natural selection to work, Julian Huxley referred to this period as the eclipse of Darwinism.[6] The natural historians that opposed Darwinian natural selection sought alternatives.

Darwin's ideas pre-dated Mendel, and he was unable to provide the evidence for heredity that Mendelian genetics did. From 1900-1918, a rivalry existed between the neo-Darwinian and Mendelian schools.

The Mendelians were interested in the mechanisms of heritability. In 1900, Carl Correns, Hugo de Vries, and Erich von Tschermak rediscovered Mendel's 1866 paper published in German, *Experiments in Plant Hybridization*. Correns and de Vries were interested in the mutation theory and von Tschermak recognized the practical use of Mendel's laws and applied them to developing new crops.[7]

In 1901, plant physiologist Hugo de Vries discovered mutations in the

Table 5.2 Not a one man show

1543 Copernicus proposed a heliocentric universe
1610 Galileo's proof of a heliocentric universe
1735 Linnaeus' *System Naturae* classification system
1745 Maupertius recognizes adaptation and fitness
1749 Buffon's 36 volumes on natural history discussed human and ape similarity, regions and climates resulting in biodiversity, and common descent
1788 Buffon revises geological timeline
1796 Cuvier's comparative anatomy, paleontology, and extinction via catastrophes
1798 Thomas Malthus' *An Essay on the Principle of Populations*
1809 Lamarck's use and disuse theory and the inheritance of acquired characteristics
1813 William Charles Wells recognizes principle of natural selection
1830 Charles Lyell observed humans do not appear in the fossil record
1844 Robert Chambers' *Vestiges of the Natural History of Creation*
1855 Alfred Wallace's biogeography
1859 Pasteur's germ theory
1866 Mendel's Laws of Classical Genetics

evening primrose that resulted in a new species. His discovery and hybridization experiments enabled quick speciation in contrast to Darwin's slow accrual of mutations with transition forms. Subsequently, the Mendelians led by William Bateson proposed that mutations drive evolution.

THE SCIENTIFIC COMMUNITY UNITED

Theodosius Dobzhansky, who worked in Thomas Hunt Morgan's lab in California in the 1930s, collaborated with population geneticists Sewall Wright, J.S.B. Haldane, and R.A. Fisher to help formulate a synthesis that reconciled the mutation, neo-Darwinism, and Mendelian schools. Fisher and Haldane were Darwinians, and Wright was interested in mutations and migration.

From Mendel's genetic experiments, we know there are two alleles for each gene, dominant (A) and recessive (a). Also, we know there are three possible genotypes (see Table 5.3), homozygous dominant (AA), heterozygote (Aa) which has twice the probability of occurring, and homozygous recessive (aa). Using the Hardy-Weinberg equation, it is possible to mathematically calculate the equilibrium of gene pool frequencies in a population

In 1908, mathematician G. H. Hardy and Wilhelm Weinberg, a physician, determined that allele frequencies in randomly breeding popula-

Table 5.3 Distribution of alleles

AA	Aa
Aa	aa

tions remain constant from generation to generation known as the Hardy-Weinberg principle. Evolution involves changes in the gene pool, and several factors are responsible for these changes. These factors include mutations, migration (movement from population to population), genetic drift (random change in allele frequency), non-random mating, and natural selection. Using the Hardy-Weinberg principle, the researchers were able to track alleles across populations over time.

While most of these factors were well understood by natural historians, genetic drift was not. When genetic drift occurs, only a small fraction of possible genes are passed on. If an event wipes out a significant percent of a population or species it leaves the future generations with limited genetic diversity. In a large population this will not have much effect; however in a small population the effect is significant. For example, an animal species driven to near extinction by reduction in habitat or hunting has a significantly reduced genetic variation, referred to as a population bottleneck.

Genetic drift can also occur if a small group breaks off from a larger population. For example, the Amish split off from Europeans and settled in the United States, and the Afrikaners split off from Europeans and settled in South Africa. If one of the individuals carries a rare allele, or version of a gene, related to a disease, that allele will have a higher frequency than it had in the larger group and result in higher frequencies of genetic diseases. Many beneficial adaptations are also eliminated, referred to as the founder effect. For example, when cattle and other species of animals are transported to other parts of the world, gene flow results. Gene flow resulting from migration out of a population results in a loss of gene pool frequencies, and migration into a population results in a gain of gene pool frequencies.

Population geneticists solved the problem of adaptive changes in populations. Natural selection is a process that alters allele frequencies. In 1942, Julian Huxley coined the integration of Mendelian genetics, Darwinism, and mutations with population genetics, as the Modern Synthesis. It unified biology and formed the basis for understanding microevolution; genetic changes at the individual level.

Haldane proposed that genetic diseases may have evolved through natural selection. Scientists have determined mutations are responsible for cystic fibrosis, sickle cell anemia, Huntington's disease, hemophilia, and color blindness. These diseases follow a Mendelian pattern of inheritance.

In 1949 William Castle and Linus Pauling identified an abnormal hemoglobin molecule in the red blood cells of patients with sickle cell anemia. The researchers discovered the change of a single amino acid. That change is an adaptation to prevent malaria, which is responsible for millions of deaths annually. The amino acid change results in an alteration to the hemoglobin molecule's shape which prevents it from carrying normal levels of oxygen. This mutation makes it more difficult for *Plasmodia*, the bacteria responsible for malaria, to infect red blood cells.

Since humans are diploid, one allele is inherited from each parent, which is either dominant or recessive. Based on Mendelian genetics, predictable variations of phenotypes can occur.

- Sickle cell anemia results in individuals homozygous for the sickle-cell allele where both parents are carriers of the recessive alleles. The sickle-cell anemia allele has a high frequency in Africa where malaria is common.
- Heterozygotes, those with one normal allele and one copy of the sickle cell allele, have the highest fitness with resistance to malaria.
- Individuals homozygous for the normal allele have the highest incidence of malaria. When bacteria destroy normal red blood cells, anemia results. Without treatment, infected red blood cells can block vessels to the brain and other organs with fatal results.

Cystic fibrosis is one of the most common diseases affecting those of European descent. Approximately 30,000 Americans who have recessive alleles from both parents have the disease. An estimated 12 million Americans have one copy of the recessive gene and are carriers. Scientists believe it is linked to a 50,000 year old mutation to prevent cholera, typhoid, or tuberculosis.

In 1990, the labs of Lap-Chee Tsui and Francis Collins discovered the mutated allele responsible for cystic fibrosis. The mutated CFTR allele, found on the human chromosome 7, encodes for a protein that transports chloride ions and water into the lungs and other tissues. One mutation is a deletion of three nucleotides that code for the amino acid phenylalanine, which appears in 70 percent of cystic fibrosis cases.[8] CFTR mutations inhibit the ability of cells to transport water into and out of cells resulting in the accumulation of mucus. As mucus accumulates in the lungs, it increases susceptibility to respiratory infections and destroys lung tissue.

Consequently, cystic fibrosis is sometimes fatal.

The mutation causing cystic fibrosis provides an evolutionary advantage. Gerald Pier of Harvard Medical School discovered the protein encoded by the normal cystic fibrosis gene is a receptor for *Salmonella typhi*, the gastrointestinal pathogen that causes typhoid fever. The mutated CFTR receptor protein provides resistance to the *Salmonella typhi* bacterium.

With cystic fibrosis, individuals acquire a mutation in the CFTR allele, which is positive selection, and inherited through Mendelian genetics, through double recessive alleles. Under the Kuhnian model of scientific revolutions, the Modern Synthesis, rather than Darwinian natural selection is revolutionary since it provides a mechanism for simple phenotypes. The Modern Synthesis which unified Mendelian genetics, Darwinian natural selection, and mutations provided a new paradigm that scientists agree upon.

THE *HOX* PARADOX

In the 1930s population geneticists explained the adaptive changes in populations, which solved the mechanism for natural selection. However, the other main question in biology at the time remained — the mechanism for speciation. The Modern Synthesis does not explain speciation for several reasons.

First, in population genetics a single locus, location of a single gene, is used. The Hardy-Weinberg principle uses selection of diploids, genes inherited from both parents, at one locus, so it does not apply to complex phenotypes. The fourteen species of finches that Darwin that observed developed from a single species of finch that reached the Galapagos Islands. Their speciation resulted from competition for food in an ecological niche within a geographic range. If allele frequencies remain constant the organism remains in stasis, which is the case with the coelacanth; however, failing to acquire adaptive mutations usually leads to extinction. If allele frequencies change that allow organisms such as Darwin's finches to adapt to a new source of food, they will acquire adaptive beaks; otherwise they may go extinct.

Secondly, Darwinian evolution accounts for changes in a population's gene pool that are gradual; the slow accrual of mutations which occur over many generations. The Modern Synthesis assumed microevolution, the change of allele frequencies from selection, mutation, genetic drift, and migration, as the only modes of evolution. The emphasis on microevolution split the scientific community. In contrast to adaptationists, who rely solely on the interaction of an organism and the environment, evolutionists argue that evolution arise from a number of interacting processes. Macroevolution, including speciation resulting from catastrophes, symbiosis, and lateral gene transfer, provide rapid changes in the genome.

According to the rules of Mendelian inheritance, the ratios of traits in offspring are predictable; however, allele frequencies in populations are not. While natural selection acts at the individual level, evolution is the result of multiple forces acting on populations. Today scientists understand evolution as the interaction of many processes with gradual and rapid changes at a number of levels.

Third, population genetics ignores embryological development. Evolution is the result of changes in development, not gene frequencies. The Modern Synthesis did not account for how organisms develop from womb to adult. To understand speciation required breakthroughs in developmental biology.

Every animal is the product of two processes: the development from an egg, and evolution from its ancestors. Ontogeny is the development of an embryo to adult, and phylogeny is the evolutionary development of an organism from ancestral to modern species. In 1866, based on similarities between fish, bird, and mammal embryos, German zoologist Ernst Haeckel proposed the theory of recapitulation to explain speciation. This theory proposed that embryonic development retold the story of morphological evolution referred to as ontogeny recapitulates phylogeny. According to Dalhousie University biologist Brian Hall:

> The belief in recapitulation of adult ancestors in the ontogenetic sequences of their descendants strangled any meaningful integration of development and evolution for almost 100 years.[9]

Today we know that Haeckel's ontogeny recapitulates phylogeny theory is incorrect. The development of an embryo does not fully repeat the evolutionary development of the species. Species that have an evolutionary relationship typically share the early stages of embryonic development and differ in later stages, which Haeckel referred to as heterochrony. If a phenotype has earlier origins, it also appears earlier in the embryo. Conversely, the loss of a phenotype will show early stages of embryonic development. Human embryos have vestiges of a tail at one point. The cerebrum, the most advanced part of the brain in humans, develops last.

Brain and head development in chimpanzees and humans is similar before birth; however humans continue their brain development and head growth several years after birth.

In a critique of Haeckel's recapitulation theory, in 1928 Karl von Baer argued animal life has four fundamental arrangements or archetypes that represent different grades of formation that are distinct in nature.[10] To explain this phenomenon, in 1942 developmental biologist Conrad Waddington introduced the concept of canalization. This is a process where mechanisms buffer a phenotype against drastic changes resulting from environmental influences. In this process, anatomical modules are programmed separately without affecting other parts of the embryo. Although canalization is works to preserve anatomical features, other forms of development are less robust.

Evolution occurs at three levels. First, changes in gene frequency which operate at the population level through mutations, selection, drift, migration, and meiotic drive. Meiotic drive is a process which causes some aggressive alleles called drivers to be over represented in the gametes formed during meiosis. The other two levels, the appearance of new characters and the appearance of new species through adaptation and radiation, require alterations in development.[11]

Understanding how the bodies different as insects and mammals are derived by the same developmental regulatory gene networks resulted in a mystery known as the *Hox* paradox.[12] Despite many similarities in gene expression exhibited in the fruit fly and the mouse, their embryos give rise to adults that are very different anatomically which questions a common ancestry.

Philosopher of biology Sahotra Sarkar observed, "In the excitement following the pursuit of the possibility that information resided in DNA sequences, alternatives were ignored."[13] One alternative is epigenetic regulation (see Figure 5.2). Activation or inactivation of an allele is one method of determining gene expression.

A second alternative is EvoDevo (evolutionary and developmental biology). Researchers in this field investigate how transcriptional regulatory systems evolve, how they contribute to phenotype, and why evolu-

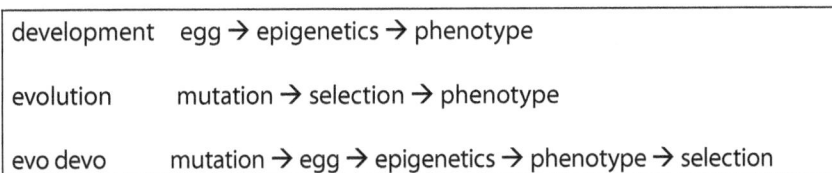

development	egg → epigenetics → phenotype
evolution	mutation → selection → phenotype
evo devo	mutation → egg → epigenetics → phenotype → selection

Figure 5.2 Epigenetic regulation of alleles and EvoDevo[14]

tion chose this path. A major breakthrough in EvoDevo occurred in 1984 when Mike Levine and Bill McGinnis discovered *Hox* genes, regulatory genes that act as master switches, that turn other genes on and off. When researchers cloned and sequenced regulatory genes, they were able to establish a link between genes and development.[15]

In *Regulating Evolution: How Gene Switches Make Life*, evolutionary biologists Sean Carroll, Nicholas Gompel, and Benjamin Prudhomme explain the basic mechanisms of how EvoDevo works.

> The expression of a gene requires transcription of DNA into mRNA and translation into a protein. Expression at the transcriptional level is regulated by non-coding DNA sequences. These non-coding DNA sequences are components of genetic switches that turn genes on and off in certain parts of the body. Sequence specific DNA binding proteins called transcription factors, which are the other components of the switch, recognize those DNA sequences referred to as enhancers. The binding of the transcription factors to the enhancers within a cell nucleus determines whether the switch and the gene are on or off in that cell.[16]

Animals share the same regulatory genes that are responsible for the development of body segments such as appendages. Although *Hox* genes are highly evolutionarily conserved in vertebrates, their interactions are not.[17] In the 1990s, when it was understood that regulatory genes acquired novel roles, this explained variable anatomical expression. By changing the order of the expression of *Hox* genes during the development of fruit flies, scientists were able to form legs instead of antennae.[18] The conservation of regulatory genes in embryos enables birds to express one combination of genes resulting in wings and feathers, while humans express another combination resulting in limbs.

In one species of *Drosophila* (fruit fly), the male has a spot on his wing that is displayed during courtship. Carroll of the University of Wisconsin, and Gompel and Prudhomme of the Institut de Biologie du Developpement in France investigated how the spot is produced by searching for enhancers that control gene expression in adjacent DNA sequences. In 2008, the researchers discovered the spotted species acquired new binding sites for transcription factors that enable the expression of a protein at high levels creating the wing spots.[19]

Genes that encode proteins that shape anatomy have multiple enhancers which allow diverse uses for the same gene. Modular regulation through mutations in enhancers allows the modification of individual body traits without changing genes or proteins. During development, shifts in inputs where *Hox* genes are expressed in embryos enable the development of different body forms, sizes and shapes. Using combinatorial logic, a series of temporal and spatial inputs from the environment determine gene expression including butterfly spots and zebra stripes.[20]

Mutations in regulatory DNA have played a major role in human evolution not only for physical traits, but also by providing protection to diseases. In addition to the major blood types A, B, and O there are also minor blood types. An example is Duffy proteins present on the surface of red blood cells that help make up receptors used by the malaria causing parasite *Plasmodium*. The protein is virtually absent from red blood cells in the population of West Africans making them resistant to infection. The West Africans have a mutation in an enhancer of the Duffy gene that knocks out the binding site for a transcription factor that activates Duffy expression in red blood cells. However, the mutations have no effects on cells in the kidneys, spleen or brain where the Duffy gene is also expressed.

According to Greg Wray, a developmental biologist at Duke University, researchers have discovered two common situations that enable mutations such as Duffy to occur. One scenario is the regulatory elements that control expression in other tissues are physically separate from those that regulate red cell expression. Another possibility is a single regulatory element controls expression in multiple tissues, but the mutation that provides protection to malaria infection only interferes with protein-DNA interactions that occur in red cells.[21]

The development of enhancers has an evolutionary advantage because most body patterning and anatomical genes are pleiotropic or used for multiple traits. Consequently, the loss of function of a gene would disrupt functions in other parts of the body. If the cell gambles with changes that do not work, it fails to generate new cells and all of its information is lost. Since genetic information is passed from generation to generation, cells must maintain a line back to the earliest cells. The loss of these genes would.

Enhancers enable the same gene to play different roles. This mechanism allowed the loss of limbs in reptiles and whales which are adaptations for different habitats and methods of locomotion. Using the mechanisms of EvoDevo, organisms can have variable gene expression resulting in necks that are broad or narrow and long or short. Scientists believe the giraffe's neck developed via *Hox* genes.

A CULTURAL REVOLUTION

So, how does Darwinian natural selection currently fit in to historical context with other scientists' work? In the March 2005 issue of the *Journal of the History of Biology*, Michael Ruse suggests the Darwinian revolution was cultural, not biological. Ruse further suggests that Darwinism was not just an alternative to the Judeo-Christian account of origins, but also used society as a tool of reform to move science forward.[22] The contributions by Darwin and others made evolution a believable theory and created a more receptive audience in the scientific community.

As expected, the reception to Darwin's theory of evolution via natural selection outside the scientific community was not without a fight. In 1860, months after the release of *Origins*, Bishop Samuel Wilberforce defended Christianity against Darwinian evolution in a formal debate at Oxford University. Thomas Huxley filled in for Darwin, who was too ill to attend. Wilberforce's main line of attack was declaring that a common ancestry with monkeys is preposterous.

The fight later moved across the pond creating a controversy in the American educational system. In 1914, evolution first appeared in American textbooks. In 1923, William Jennings Bryan led a successful campaign outlawing the teaching of evolution in public schools in Oklahoma and Florida.

A similar law passed in Tennessee in 1925, the Butler Law, which made it illegal to teach any theory that denies the Bible's creation story in state funded schools. The "Scopes Monkey Trial" became a test case for teaching evolution in American public schools. In the trial, prosecutor William Jennings Bryan defended the Bible against ACLU criminal defense attorney Clarence Darrow representing John Scopes, a high school teacher in Tennessee. The court found Scopes guilty of teaching evolution and fined him. Subsequently, the publishers of biology textbooks conformed to the Tennessee state ruling.

Based on the Establishment Clause, a provision of the First Amendment of the U.S. Constitution that prevents laws establishing a state religion or forcing the belief of any religion, later court rulings overturned the *Tennessee v. John Thomas Scopes* decision, repealed the state law, and allowed the teaching of evolution. In 1950 Pope Pius XII wrote in *Humani generis* that the account of biblical creationism in Genesis was not a literal, rather a metaphorical, source of human origins.[23]

In the 1980s, creationists led by retired UC Berkeley law professor Phillip Johnson began selling creationism under the new name of intelligent design and attacking evolution. In *Edwards v. Aguillard* (1987), the court determined that the definition of creationism and intelligent design are identical. Creationism appeared 150 times and was crossed out and replaced with intelligent design in a revised draft of *Of Pandas and People*.[24] The Supreme Court ruled that the Louisiana school board's intentions were not to teach all scientific theories but rather to discredit evolution which violated the Establishment Clause. In spite of the ruling, Johnson's *Darwinism on Trial* (1991) became the manual for the intelligent design movement. A conference held at Southern Methodist University in 1992 brought together creationists to discuss strategies to continue opposing evolution.

A series of creationist books including Michael Denton's *Evolution: A Theory in Crisis* (1986), and Johnson's *Defeating Darwinism by Opening Minds* (1997) and *Reason in the Balance* (1998) continued to critique evolution on the basis that that Darwinian evolution doesn't adequately explain speciation. These creationist books argued that too many gaps exist in the fossil evidence. If evolution occurred, paleontologists would have discovered more preserved intermediate stages between lower and higher forms of life in the fossil record.

In *Darwin's Black Box* (1996), Michael Behe, a Fellow at the Discovery Institute, a Christian think tank in Seattle, argues irreducibly complex motors and circuits found in cells are not reducible to physics and chemistry.[25] In Behe's testimony at the *Kitzmiller v. Dover Area School District* (2005) trial he asked for an example of a new species formed by the accumulation of mutations. In *The Edge of Evolution* (2007), Behe, who is also a Professor of Biochemistry at Lehigh University, downplays the significance of Carroll's EvoDevo findings saying he provides no specifics on how particular structures would evolve by random mutations and natural selection.[26]

In 1998, The Discovery Institute formally drafted the details of the movement into a document called the Wedge Strategy. The document outlines a five year plan to make intelligent design an alternative to Darwinism. Leaders of the movement planned to accomplish this strategy through a public relations campaign attacking Darwinism at the weakest points to make it controversial.[27] The long term goals were outlined in a twenty year plan to make intelligent design the dominant perspective in science as well as our political and cultural life.

The movement targets the public and media with publications and appearances, and lobbies educators and policy makers for teaching intelligent design in public schools. In 2001, Senator Rick Santorum introduced the Santorum Amendment, an amendment to the No Child Left Behind Act, to Teach the Controversy which promotes the teaching of intelligent design and emphasizes the weaknesses of evolution in public school classes.

In 2004, the school board of a small town in Pennsylvania voted to have a requirement that biology teachers teach evolution as a theory and propose intelligent design a supernatural designer. In *Kitzmiller v. Dover Area School District* (2005), the plaintiffs successfully argued that intelligent design is a form of creationism and that the school board policy violates the Establishment Clause of the First Amendment. In addition, the judge required that the Dover school district pay the six-figure legal fees.

In *Darwin's Dangerous Idea* (1995), philosopher Daniel Dennett points out that the intelligent design movement inaccurately describes science and creates a false perception that evolution is a theory in crisis rather than contributing to education. The Intelligent Design movement is currently left without legal or scientific standing.

Numerous court decisions have forced the movement to give up on the idea that creationism and evolution are science based alternatives to human origins taught in high school science curricula. Consequently, creationists have developed a new strategy. The Discovery Institute is now providing a template to legislators to file academic freedom bills which promote teaching the strengths and weaknesses of evolution in the science curriculum.[28]

Current polling reveals when given a choice of either the descent of man or creationism for human origins, most Americans do not believe in evolution. So, it is not accurate to say that evolution is revolutionary in the way most Americans view human origins. However, more respondents are more likely to believe in evolution if the polling questions are related to the co-existence of creationism and evolution.[29] In two recent books, Francis Collins' *The Language of God* and E. O. Wilson's *The Creation*, the noted scientists argue that evolution and intelligent design are compatible.

In contrast, according to the NIH and the AAAS, scientists accept evolution and reject intelligent design by an overwhelming majority. The National Association of Biology Teachers has endorsed evolution quoting evolutionary biologist Theodosius Dobzhansky, "Nothing makes sense except in the light of evolution."[30]

Chapter 6

Four Waves

For over two millennium, natural historians based their explanations to nature's mysteries, the creation of the universe and its life forms, on vitalism. The current scientific understanding of the creation of the universe, the earth, and its life forms, provides a revised timeline (see Table 6.1).

In contrast to the creation theory, scientists now believe the solar system formed as part of the Big Bang 13.7 billion years ago. Based on the radioactive decay in isotopes of Uranium in the mineral Zircon found in meteorites, scientists believe the earth formed approximately 4.54 billion years ago.

Aristotle's purpose, LaMarck's adaptive forces, and Darwin's gemmules were mechanisms proposed by natural historians to the puzzles of heredity and variations in life forms. Four waves of discoveries in molecu-

Table 6.1 Timeline for the universe and its life forms

• 13.7 billion years ago The Big Bang
• 4.54 billion years ago the formation of the earth
• 3.5-4.5 billion years ago self-reproducing simple cells
• 2.8-3.3 billion years ago Archaen Expansion
• 3 billion years ago prokaryotes
• 2 billion years ago eukaryotes
• 580 million years ago the Cambrian explosion
• 200 million years ago mammals
• 6 million years ago chimpanzees and humans split
• 350,000 years ago Neanderthals
• 200,000 years ago *Homo sapiens*
• 50,000 years ago migration Out of Africa

lar genetics would provide alternative views to vitalism.

In 1800, Karl Friedrich Burdach coined *"biologie,"* or as we know it today "biology." However, the rise of biology as a profession did not occur until a lifespan later. In 1902, Archibald Garrod proposed the concept of a unit of heredity. Gregor Mendel referred to this unit as a "factor." In 1905, Danish botanist Wilhelm Johannsen coined the word "gene" to differentiate from Darwin's pangenes and theory of pangenesis. However, scientists did not understand the role of genes and the actual genetic material remained a mystery for a number of decades.

The first two waves of discoveries led to the rise of the genetic program. The genetic program is composed of two dogmas: the reductionism of genetics to a physical unit and genetic determinism with DNA as a blueprint. This led to a gene centered view of evolution.

Two more waves of discoveries would follow, which would lead to the descent of the genetic program. Along with improvements in dating methods for fossil evidence, deciphering the human genome has also provided insight into our evolutionary history. Sequencing and comparing genomes has enabled scientists to provide a strong argument for common descent that was not possible in Darwin's lifetime.

In *The Century of the Gene*, Evelyn Fox Keller notes scientists once believed that decoding the message in DNA sequences would provide an understanding of the program that makes an organism. However, "structural genomics has given us the insight we needed to confront our own hubris, insight that could illuminate the limits of the vision with which we began."[1]

THE SEARCH FOR A UNIT OF HEREDITY

In the 1870s, a series of educational and economic reforms took place in Germany promoting private initiative.[2] As part of the reforms the German government incorporated biological research in its universities. Consequently, what was once a hobby for the well-to-do natural historians was elevated to a respected profession.

In the search for a unit of heredity, European scientists initiated the first wave of discoveries in molecular biology. In 1870, while working with used bandages, Friedrich Miescher isolated substances called nuclein from human pus which were later renamed nucleic acids. In 1879, after observing the exchange of genetic material in stained bodies, Walter Flemming discovered chromatin, which Wilhelm Waldeyer later renamed chromosomes, Greek for colored body.

The invention of the compound microscope enabled Robert Hooke to discover cells. In 1858, Rudolph Virchow proposed that cells arise from preexisting cells that divide, and these specialized cells then develop into organs. In 1887, August Weismann discovered that germ cells reproduce by meiosis. Weismann proposed the chromosome theory of heredity; however, he was unable to provide evidence of a physical unit of inheritance.

In 1902, while studying grasshopper cells, Walter Sutton observed that during meiosis chromosomes pairs split, making two daughter cells. By exchanging DNA during meiosis, chromosomes increase genetic variation; he provided the first proof that chromosomes carry hereditary material. Sutton wrote:

> Many points were discovered which strongly indicate that the position of the bivalent chromosomes in the equatorial plate of the reducing division is purely a matter of chance — that is, that any chromosome pair may lie with maternal or paternal chromatids indifferently toward either pole irrespective if the positions of other pairs — and hence that a large number of different combinations of maternal and paternal chromosomes are possible in the mature germ-products of the individual.[3]

In 1910, Thomas Hunt Morgan of Columbia University revived Weismann's chromosomal theory of heredity. Research in his lab, known as the "Fly Room," led to production of a cross bred mutant fruit fly, *Drosophilia melamogaster*, with solid white eyes. The wild-type or most common phenotype for the fly is red eyes. He discovered solid white eyes appeared only in males as a Mendelian sex-linked recessive trait.

Morgan established that genes are located on specific chromosomes, confirming Walter Sutton's prediction. Morgan proposed crossing-over, the exchange of genetic material between chromosomes. Barbara McClintock and Harriet Creighton later confirmed Morgan's finding. Biologists credit Morgan with integrating Mendel's laws and the chromosome theory of heredity together into a believable concept.

We know from Mendel's Law of Segregation that each gene separates into different cells during the formation of gametes (egg and sperm). Each germ cell becomes haploid, having one copy of each chromosome, and the diploid number is restored at fertilization with copies of chromosomes from both parents. Normal human germ cells nuclei contain two sets of twenty-three chromosomes with one set inherited from each parent.

Table 6.2 The first wave of discoveries in molecular biology

1870 Meischer discovered nuclein
1882 Walter Fleming discovered mitosis
1887 Weismann discovered germ cells reproduce by meiosis.
1888 Wilhelm Waldeyer coins chromosome
1902 Sutton proposed the chromosomal theory of heredity
1902 Archibald Garrod proposed a unit of heredity produces specific proteins
1905 Danish biologist Johannsen coins gene
1910 Thomas Hunt Morgan established genes are located on chromosomes

Chromosomes one through twenty two are referred to as autosomes, and the twenty-third set is either the female XX or the male XY sex chromosome.

Humans are diploid and offspring may inherit the same allele or version of a gene from each parent. For example, humans may inherit the same alleles that determine either A, B, or O blood types. These offspring are homozygous AA, BB, and OO. If the offspring inherit different alleles, they are heterozygous AB, AO, or BO. In humans, A and B blood types are co-dominant, so A and B are expressed, and O is recessive. As a result, A and B blood types are the most common.

THE RELATIONSHIP BETWEEN GENES AND PROTEINS

Initially, American scientists went to Germany for biological training. Then, patterned after the German system, universities in the United States began adopting biological research in their medical and graduate schools. Johns Hopkins was the first in 1876. Harvard, the University of Chicago, and Columbia University soon followed. Researchers from these schools became pioneers in a second wave of discoveries in the fields of molecular biology and genetics, establishing the relationship between genes and proteins.

Following the Spanish flu epidemic of 1919, Frederick Griffith in conjunction with the British Ministry of Health worked on developing a vaccine for lobar pneumonia. Griffith injected two strains of the *Pneumococcus* bacteria into mice. The virulent S form, which has a polysaccharide capsule, evades the host's immune system was lethal, and the non-virulent R form is recognized by the host's immune system. When living harmless R cells were combined with heat-killed S cells the combination was lethal when injected. In 1928, Griffith discovered that by mixing the two forms, the cells recovered. Mice had genetic material transferred from the dead virulent form to the live non-virulent form. The next goal was to find the

Table 6.3 Timeline for understanding the relationship between genes and proteins

1902 Garrod proposed a link between a unit of heredity and proteins
1941 Beadle and Tatum propose the one gene-one protein theory
1944 The Transforming Principle by Avery, MacLeod, and McCarty
1950 Chargaff's rules on nucleotide pairing
1952 Franklin and Gosling's X-ray crystallography image photograph 51
1953 Watson and Crick established the double helical structure of DNA
1958 Crick and Masov proposed the central dogma

S form's transforming factor. At the time, some scientists proposed genes, while others believed it was proteins.

George Beadle and Edward Tatum aimed to prove experimentally Garrod's prediction that a unit of heredity is responsible for producing specific proteins by creating single gene mutations in bread mold that affected specific enzymes. In 1941, based on their experiments, they proposed the one gene-one protein concept. If this theory proved accurate, based on an estimated 100,000-150,000 proteins, this would indicate a similar number of protein coding genes exist.

Using a process of elimination, researchers at the Rockefeller Institute for Medical Research tested both theories by labeling DNA and proteins with radioactive isotopes. Within a controlled environment, using enzymes to inactivate genes genetic information didn't transform. However, using protease enzymes to remove proteins, the R strain bacteria maintained the ability to transfer genetic material.

By demonstrating that a transforming or inherited factor in the bacteria is capable of transferring from the non-virulent to the virulent form, Oswald T. Avery and colleagues Maclyn McCarty and Colin MacLeod became the first to demonstrate genes are composed of DNA. The conclusion section of their 1944 paper citing the discovery includes a single sentence.

> The evidence presented supports the belief that a nucleic acid of the deoxyribose type is the fundamental unit of the transforming principle of *pneumococcus* Type III. [4]

Many scientists believe this work is deserving of a Nobel Prize. The discovery was influential not only in the development of molecular biology, but in genetic engineering as well. Dr. Joshua Lederberg, a Nobel Prize-winner and a former head of Rockefeller University noted:

> It was the pivotal discovery of 20th-century biology. Nobel Prize nominations are complicated, but everybody includ-

ing the Nobel Committee will acknowledge that this was
their most significant failure. There must be 20 to 25 prizes
that have been awarded for work that depends on the team's
seminal paper.[5]

Martin Hewlett observed that Avery's work did not have immediate
impact because of Phoebus Levene's model of DNA structure. Levene had
incorrectly proposed DNA molecules contain equal amounts of the four
nucleotides — adenine (A), guanine (G), thymine (T), and cytosine (C). He
also incorrectly believed DNA is too simple to contain genetic information.[6]

In *What is Life?* (1944), Erwin Schrodinger challenged physicists to
examine the structure of genes. Max DelBruck and his colleagues at the
Phage Group; Alfred Hershey, Martha Chase, Max DelBruck, Salvidor
Luria, and Rosiland Franklin responded.[7] In the 1940s, The Phage Group
began studying bacteriophages, viruses that infect bacteria, and discovered that bacteriophages were able to attack only one type of bacteria. In
1952 using radioactive tracers in proteins and DNA, Alfred Hershey and
Martha Chase confirmed that DNA encodes hereditary information. The
scientific community finally accepted that genes are composed of DNA.

THE STRUCTURE OF DNA

With Avery's breakthrough that genes are composed of DNA, the next
crucial step to understanding molecular genetics was a better understanding of DNA's structure. Because of, Erwin Chargaff of Columbia University changed the focus of his laboratory work from lipoproteins to nucleic
acids and the structure of DNA. In 1950, Chargaff observed that the adenine content of DNA is the same as thymine and that the guanine content
is equal to cytosine. Complementary base pairing in chromosome replica-
tion is now known as Chargaff's rules.

In 1950, James Watson was as a post-doctoral researcher in Copen-
hagen supported by the Merck Foundation with a focus on nucleic acid
chemistry and bacterial viruses. There he met Maurice Wilkins, a bio-
physicist at King's College London, at a conference which included a presentation of a paper on X-ray diffraction and crystallography. Soon after,
Watson moved his fellowship to Lawrence Bragg's Cavendish Laboratory
at Cambridge University. There he met doctoral student Francis Crick.

Also at King's College London was Rosalind Franklin who was improving upon X-ray diffraction techniques to better understand the structure of DNA. After bombarding DNA with X-rays and analyzing how it
scattered the radiation, Franklin and Wilkins calculated the distances be-

tween groups of atoms in the molecule. The scattering patterns showed that three distances were repeated many times in the secondary structure, and from this they inferred that DNA has a helical shape.[8]

The discovery that DNA has a helical shape led several scientists to propose models for the structure of DNA including Watson and also Linus Pauling and Robert Corey of the California Institute of Technology. Watson initially proposed a triple helical structure of DNA using low quality photos. After viewing the presentation of Watson's models and the photo from which the model was based, Franklin observed that the calculation of hydration had led to misleading density data.[9] The amount of water used in DNA samples leads to different conformations.

In 1952, using the most advanced X-ray crystallography at the time, Franklin and her graduate student Raymond Gosling produced X-ray photograph 51. In 1953, based on photograph 51 with DNA with strands wound together, Chargaff's complimentary base pairing, and discoveries from their own research, Watson and Crick developed a model with a double helix. The three dimensional model includes two long polymers consisting of nucleotides with back¬bones made of sugar molecules and phosphate groups.

Crick, Watson, and Wilkins received the Nobel Prize for Physiology and Medicine in 1962 for their discovery. Although Franklin's X-ray diffraction techniques contributed to the discovery, she was deceased and the prize is limited to three people. Ironically, Wilkin's received the Nobel Prize for his lower quality X-ray diffraction pictures.

THE CENTRAL DOGMA

Researchers now placed more emphasis on the mechanisms that would enable the production of proteins from DNA. In 1955, Crick proposed The Adapter Hypothesis, which stated that a molecule reads the nucleic acid code and transfers the information from mRNA into the amino acid sequence of a protein. In 1956, Paul Berg of Stanford University successfully identified these adapters, now known as transfer RNAs (tRNAs), which are able to bridge the codes of nucleic acids and amino acids. Researchers later found that more than one tRNA exists for each amino acid due to redundant codons.

In 1956, Oak Ridge National Laboratories researchers Elliot Volkin and Lazarus Astrachan infected *E. coli* cells with a bacteriophage. They discovered the bacteriophage turns off the bacterial cell's ability for making its own proteins and instructs the cells to make proteins characteristic of the virus. They called the molecule with instructions a DNA-like RNA, because it transports information from genes in the nucleus to the cytoplasm. Jacob and Monad later designated these molecules as messenger RNA (mRNA).

Scientists had now discovered two types of genetic information, genes and mRNAs. In 1956, Crick and George Gamov, a Russian physicist, formulated a linear relationship between DNA and proteins referred to as the central dogma (see Fig. 6.1). This genetic deterministic relationship, however, does not account for environmental factors.

The first step, transcription is the synthesis of an mRNA molecule. A promoter region facilitates the initiation of transcription. In humans, non-protein coding DNA or intronic regions are spliced out by splicesosomes, and exonic regions or protein coding DNA are pieced together to make mRNA. In eukaryotic cells, the transcription step is necessary because the genetic material in the nucleus is physically separated by a nuclear membrane from the cytoplasm, the site of protein synthesis in the cell.

The mRNA then moves to the ribosome for the second step called translation referred to as protein synthesis. The mRNAs are translated into proteins which have specific three-dimensional structures and functions. Crick proposed that similar to DNA and nucleic acids the specificity of proteins resides in the sequences of amino acids. According to Crick's 1958 sequence hypothesis, the amino acid sequence determines protein shape which in turn determines its function. However, the mechanism that translated mRNAs into proteins remained a mystery.

In 1966, Johan Matthaei, Marshall Nirenberg, and H. Gobind Khorana led independent teams performing experiments that solved the mystery of the genetic code. They discovered during translation, ribosomes read mRNA codons, sets of three nucleotides that specify each of the twenty protein forming amino acids. From four nucleotides, sixty-four (4x4x4) possible combinations or triplets are possible.

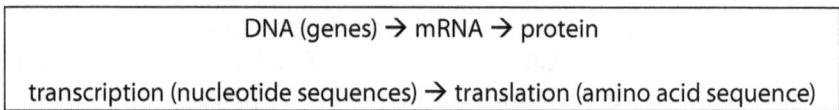

DNA (genes) → mRNA → protein
transcription (nucleotide sequences) → translation (amino acid sequence)

Figure 6.1 The Central Dogma

Table 6.4 Discoveries leading to the genetic code

1955 Crick proposed the adapter hypothesis (tRNA)
1956 Volkin and Astrachan discovered mRNA
1958 Crick proposed the sequence hypothesis
1961 Monad and Jacob discovered triplet codons
1966 Matthaei, Nirenberg, and Khorana solve the genetic code

JUMPING GENES

As a graduate student Barbara McClintock identified maize's (*Zea mays*) ten chromosomes. She then took the ten linkage groups of maize, all the genes on a single chromosome, and matched them with a specific chromosome. She observed insertions, deletions, and translocations in the DNA of the next generation. The discovery of translocations, breaks in chromosomes, that lead to transposons, the rearrangement of DNA in the same chromosome or from one chromosome to another, inspired McClintock to further investigate the underlying mechanism.

In 1944, while a researcher at Cold Spring Harbor Laboratory, she investigated the mosaic color patterns of maize. She discovered two dominant and interactive genetic loci, the specific location on the chromosome that causes it to break and change position. Although McClintock had discovered transposons, sometimes referred to as jumping genes, and transposition in the 1940s, her discoveries did not immediately gain acceptance in the scientific community.

With new technology, scientists were later able to clone the controlling units and discovered the enzyme transposase which catalyzes the movement of DNA by cut and paste. Finding the gene for the DNA cutting enzyme and transposons in other genomes confirmed McClintock's earlier discovery. McClintock would later receive the Nobel Prize for Physiology or Medicine in 1983. Historians speculate that because her theory was so radical at the time, she failed to receive earlier recognition.

THE WAR ON CANCER

As part of the War on Cancer, on December 23, 1971 Richard Nixon signed the National Cancer Act into law. This act boosted research by the National Cancer Institute and provided a challenge to eliminate the disease by 2015. Subsequently, in the 1970s and 1980s, a third wave of discoveries in molecular biology took place. Of the numerous factors that lead to cancer, it was advances in research on bacteria and viruses, and a better under¬standing the transfer of genetic material between organisms that would catch the attention of the Nobel Prize Committee.

In the 1960s, Tomas Lindahl, a Swedish cancer researcher, identified the family of mammalian DNA ligase enzymes and their specific role for joining DNA fragments together by covalent bonding. In 1970, Hamilton Smith, Werner Arber, and Daniel Nathans discovered restriction enzymes which can break DNA fragments at specific sites into 15,000-20,000 base pairs. Their discovery of restriction enzymes earned the Nobel Prize in

Physiology or Medicine in 1978. These discoveries provided a better understanding of how bacteria exchange genetic information through cutting and pasting DNA fragments.

Viruses are parasites that have the ability to use its host DNA to replicate. In 1970, while researching tumor viruses which are capable of transforming normal cells into cancerous cells, Howard Temin, Renato Dulbecco, and David Baltimore simultaneously performed independent experiments leading to the discovery of a restriction enzyme called reverse transcriptase. It is used by viruses which have RNA nucleotides to enter a host's genome. Viruses also use ligase enzymes to paste DNA fragments together.

In 1972, Paul Berg combined the DNA of SV40 monkey virus, which causes cancer in some animals, with *E. coli* via a bacteriophage, a virus that naturally infects *E. coli*, to form recombinant DNA. Berg shared the 1980 Nobel Prize in Chemistry for his advances in recombinant DNA.

In 1972, while presenting papers in Hawaii at a conference on bacterial plasmids, a serendipitous meeting of two researchers took place. Stanley Cohen, a Stanford University professor, researched ways to isolate specific genes in antibiotic carrying plasmids, small pieces of circular DNA which exist separate from the cell's main DNA and can replicate independently, and clone them individually through introducing them to *E. coli* bacteria. Herbert Boyer, from the University of California at San Francisco, discovered a restriction enzyme that cut DNA strands at specific DNA sequences producing the 'sticky ends' of DNA segments that bind together.

For researchers who want to clone a gene or to produce DNA fragments for research the first step is to isolate the DNA using restriction enzymes which recognize specific regions on the DNA molecule. Then, the isolated gene is inserted into another organism to produce proteins. The ability to transplant genes from one species to another resulted in the capability of manufacturing pharmaceutical drugs.

In 1973, Cohen and Boyer perfected genetic engineering techniques to cut and paste DNA and reproduce new DNA in bacteria. The researchers discovered it was possible to take DNA from one organism, recombine it *in vitro* with DNA of another organism, and then reintroduce it into the host. They inserted a plasmid with toad DNA into *E. coli* bacteria host, creating a transgenic organism.

These discoveries not only improved the understanding of the mechanisms for altering DNA, but would lead to genetic engineering which was instrumental in developing the biotechnology industry. Genetic engineering also referred to as recombinant DNA technology and DNA cloning has led to life saving genetically engineered drugs and genetically modified crops.

Table 6.5 Nixon's War on Cancer led to a third wave of discoveries

1940s McClintock discovered transposons
1960s Lindahl discovered DNA ligase enzymes
1970 Smith, Arber, and Nathans discovered restriction enzymes
1970 Temin, Dulbecco, and Baltimore discovered reverse transcriptase
1972 Berg's recombinant DNA molecule
1973 Cohen and Boyer discover and patent the genetic engineering process

JUST THE BEGINNING

Following the completion of the final drafts of the public and private human genomes in 2003, several prominent scientists described the accomplishment as a metaphor to the periodic table of elements. In testimony before Congress, J. Craig Venter, the chief scientist of the private human genome project referred to the acquisition of the sequence of the human genome as "just the beginning."[10]

Researchers are now investigating a new checklist of goals. With the human genome sequenced, the genetic parts list and their interrelationships and interactions are the new focus to understanding evolutionary relationships and complex diseases.

In 2003, Francis Collins and colleagues at the U.S. National Genome Research Institute prepared a report for the vision of future medical research.[11] The report lists challenges based on input from scientists and participants from workshops for addressing translating genome based knowledge into biological sciences, health benefits, and their impact on society. Among these challenges are to identify and better understand:

1. the structural and functional components of the human genome
2. genetic networks, protein pathways, and gene products
3. human traits and behaviors such as cognition and sexual orientation
4. ways in which DNA can encode information including non-genetic factors, gene-environment interactions, and epigenetics
5. evolutionary variation across species
6. genetic contributions to disease and drug response
7. develop policy options for the use of genomics
8. and to use genomic information in clinical settings, research, and education

DECIPHERING THE GENOME

Currently, scientists are in the fourth wave of discoveries spurred by the data provided by the Human Genome Project which is further unraveling of the genetic program. Shortly following the completion of the Human Genome Project, in a number of press releases Craig Venter reflected on how our understanding of DNA has changed in relation to heredity, evolution, and diseases.

> Since the June 26, 2000 announcement our understanding of the human genome has changed in the most fundamental ways. The finding that we have far fewer genes than expected suggests that environmental influences play a greater role in our development than was previously thought. The small number of genes—30,000 (later revised to 24,000) instead of 140,000—supports the notion that we are not hard-wired. We now know that the environment acting on these biological steps may be the key in making us what we are. Likewise the remarkably small numbers of genetic variations that occur in genes again suggest a significant role for environmental influences in developing each of our uniqueness.
>
> More than two thirds of these are alternative splicing genes, which code for more than one protein, and sometimes many more. We now know that the notion that one gene leads to one protein and perhaps one disease is false. One gene leads to many different products. Those products—proteins—can change dramatically after they are produced. We know that regions of the genome that are not genes may be the key to the complexity we see in humans.
>
> There are two fallacies to be avoided, determinism, the idea that all characteristics of a person are hard-wired by the genome; and reductionism, that now the human sequence is completely known, it is just a matter of time before our understanding of gene functions and interactions will provide a complete causal description of human variability.[12]

NON-CODING DNA

Perhaps the most startling discovery from analyzing the human genome was that a very small percentage of the roughly three billion nucleotide bases in the human genome encode instructions for the synthesis of proteins. Now, discovering which parts of a stretch of DNA belong to a gene means distinguishing them from the other 98 percent or so for which most of have no known function.

With cDNA (complimentary DNA or DNA synthesized from a RNA template) technology and removing the intronic regions in DNA clones, why do we not know the exact number of genes? In a matter of several years the estimate for the number of protein coding genes has been reduced from over 100,000 to roughly 22,000.

With the current estimate roughly at 22,000 genes, where is the information for over 100,000 proteins? These estimated 22,000 protein coding genes do not include microRNAs, transfer RNAs, ribosomal RNAs, or genes with short open reading frames, protein coding regions that are not interrupted by stop codons.[13] Also, alternative splicing variants and expressed cDNA under different environmental conditions do not appear.

The mechanism for proceeding to protein synthesis with large amounts of non-coding DNA requires an additional step. In humans, DNA transcribes into another type of nucleic acid, which is RNA, before used in protein synthesis. Alternative splicing occurs when enzymes called spliceosomes, a protein-RNA complex, excise the extra nucleotides and the remaining coding pieces are spliced together coding for multiple mRNA transcripts resulting in the production of more than one protein. Alternative splicing produces different mRNA transcripts from the same DNA. Alternative splicing demonstrates that genes are an intron in one type of cell and an exon in another.

In 1977, Richard Roberts and Phil Sharp discovered eukaryotic genes contain interruptions or non-protein coding DNA called intronic regions for which they won the 1993 Nobel Prize in Medicine or Physiology. They used an adenovirus, the common cold virus, that reproduces using eukaryotic cells. Earlier research performed on bacteria did not have introns, so, mRNA matched the stretches of DNA encoding it. Using restriction enzymes, they discovered the mRNA molecules were shorter than the corresponding human genes. After RNA is transcribed from DNA, the introns are cut out before mRNA is translated into proteins.

Does having less non-coding evolutionary advantage? According to Duke University cell biologist Bruce Niklas, *Drisophila* gets rid of excess DNA to have slim genomes for rapid reproduction; they have selection against having extra DNA. Humans developed spliceosomes that prevent introns from transferring to mRNAs.

Why do humans have so much non-coding DNA? Walter Gilbert proposed the evolution of new genes has proceeded by recombination and the exclusion of exons. According to the exon shuffling theory, introns act as spacers where breaks for genetic recombination occur. The organization of genes into exons separated by introns may permit rapid evolution of protein coding sequences for adaptations to environmental conditions and for new gene function. Introns could provide non-coding targets for recombination, which would then give novel combinations of exons. Exon shuffling through DNA recombinations and DNA insertions within introns can occur readily whereas in exons they must not disrupt open reading frames, DNA sequences associated with coding for proteins.

Researchers have discovered the functions of some of the introns, referred to as "junk DNA" by those that believe that protein coding DNA solely accounts for phenotypes. This non-coding DNA is classified into several categories.

REPETITIVE SEQUENCES

A large portion of non-coding DNA consists of repetitive sequences. Scientists at Celera Genomics estimated 35 percent of the human genome is composed of various classes of repeats, with chromosome 19 having the highest, roughly 57 percent.[14]

Researchers suspect some non coding DNA may have an important function, because of one particular sequence repeat called Alu, which is between 200 and 300 bases long, is clustered around genes. Alu has relatively high number of A-T dinucleotides. In 1984, Sir Alex Jeffreys, a British geneticist, discovered minisatellite DNA located near genes by accident. These short repeats are variable number tandem repeats (VNTRs) distinct in populations and are useful for forensic identification. One common dinucleotide repeat is CACACA.

PARASITIC DNA

Scientists estimate 40-50 percent is parasitic DNA originating from viral retro-transposons and bacterial transposons. Retrotransposons copy themselves and paste their instructions in the host genome in multiple places. Human endogenous retroviruses (HERVs) represent the remnants of ancestral retroviral infections that became fixed in the human germ line. These retroviruses make up as much as 8 percent of the human genome.[15]

Endogenous retroviruses contribute to the evolution of the host genome and are also associated with diseases. When a retroviral particle invades a host cell, it becomes a provirus. For example, after the HIV provirus infects humans, it passively replicates along with the host genome. Changes in health or environmental conditions may activate pro-viruses including HIV.

Analysis of the human genome reveals hundreds of human genes were obtained by lateral gene transfer from bacteria at some point in the vertebrate lineage. When T. Cavalier-Smith and Ford Doolittle compared DNA sequences that code for an enzyme from diverse organisms, they discovered the sequence from *E. coli* bacteria is more similar to the sequence found in animals than to the sequence found in other bacteria. Bacteria have several ways of transmitting DNA between unrelated organisms including transformation, transduction, and conjugation. Similar to transposons, using enzymes bacteria can excise and reinsert DNA in different locations.

Lateral transfer sequences, transposons, and pathogenicity islands (virulence genes from microbes) have created variations in the percentage of C-G content in the human genome called isochores. Stretches of up to 30,000 C-G bases repeating over and over often occur adjacent to gene-rich areas, forming a barrier between the genes and the non-coding DNA. The dinucleotide C-G is under represented in mammalian genomes. This is the result of the methylation of cytosines on both strands of DNA. Cytosines are susceptible to deamination, the removal of an amine group, which causes a mutation to thymidine. Some areas have C-G islands of unmethylated DNA which are located close to genes. These C-G islands are believed to help regulate gene activity are used as an approach to gene identification.

REGULATORY GENES

In 1948, Barbara McClintock proposed that regulatory genes act as controlling units and act as a mechanism for transposition. Scientists later made progress in understanding how specific genes are activated or silenced by regulatory genes, enhancers, promoters, and transcription factors. In 1959, Jacques Monad and Francois Jacob of the Pasteur Institute discovered a mechanism for the environmental control of protein production for which they received the Nobel Prize for Physiology or Medicine in 1965. They discovered the first gene circuit, a set of genes that help *E. coli* digest lactose. Monad and Jacob referenced McClintock's work in their paper describing the Operon Model of the genetic regulation of the lactose operon (*lac* operon).[16]

According to the Operon Model, genes are activated or turned on or off by an operator gene. An operon is a cluster of genes that is transcribed as a single mRNA which links the expression of regulatory elements and structural genes with regulatory genes elsewhere in the genome. Monad and Jacob discovered regulatory networks that include regulatory genes under the control of the same promoter that code for proteins and RNA molecules that are required to regulate the expression of other genes.

In *E. coli* bacteria, the *lac* operon regulates gene expression with one regulatory gene and three structural genes. The structural genes are involved with metabolism, and the regulatory gene is the repressor of the *lac* operon. A repressor molecule can bind to an operator of an operon to terminate transcription or the promoter region to initiate transcription.

E. coli bacterial cells will not use lactose until they consume all the glucose. So, the activation of the *lac* operon occurs when lactose is the only source of food preventing the waste of energy for the production of unnecessary proteins. The deactivation of the *lac* operon occurs both when glucose is available and lactose is not, and when both glucose and lactose are present.

Monad and Jacob demonstrated a cybernetic relationship between an organism and the environment. Controlling elements for the *lac* operon provide direct and indirect control by feedback and feedforward loops with transcription factors regulating each other. Understanding how genes are turned on and patterns of gene expression, not genes determine a cell type. We now better understand how genes are acted upon, so it is necessary to understand the environment in which cells or a species exists.

PSEUDOGENES

Pseudogenes are similar to normal genes, but are altered slightly possibly with loss of function mutations so that they are not expressed. Many of these non-functional coding sequences have too many accumulated mutations to function and produce proteins. Others may lack a promoter or have stop codons disabled through frameshift mutations. Some are thought to have originated from reverse transcribed mRNAs and reinserted into the genome since they do not contain the intronic regions that are found in the original functional gene.[17]

SMALL INTERFERING RNAS

Scientists have discovered some DNA is transcribed into RNA, but not into proteins. DNA codes for RNA sequences that alters gene expression, referred to as small interfering RNAs. These RNA segments regulate gene expression and allow cells to respond to developmental and environmental pressures.[18]

In 1993, Harvard researchers Rosalind Lee, Rhonda Feinbaum, and Victor Ambros discovered microRNAs (miRNAs) in the nematode *C. elegans*. These miRNAs are a type of small interfering RNA consisting of roughly 20-25 nucleotides. They are encoded by genes and bind to RNAs and degrade them before they are translated into proteins. Researchers attempting to engineer petunias to produce different colors by adding extra copies of genes responsible for pigments, but surprisingly the flowers lacked pigment. The extra copies were suppressing the activity of the original gene. Researchers subsequently found small RNAs present in the plants with suppressed genes.

In 1998, Andrew Fire and Craig Mello discovered a second category of small interfering RNAs, referred to as RNA interference (RNAi) which are segments of double stranded RNA for which they received the 2006 Nobel Prize in Physiology or Medicine. Scientists believe the major role for RNAi in eukaryotic organisms is a gene blocking procedure which protects the genome against exogenous viruses and endogenous transposable elements.[19] A nucleotide sequence is engineered which is complementary to the mRNA sequence produced by a gain of function mutation that prevents translation of the gene. The double stranded RNA is chopped into short sequences and used to destroy single stranded mRNA which prevents invading viruses.

Applying this knowledge of antisense technology, scientists genetically modified the Flavr Savr tomato deactivating the mechanism that causes softening in tomatoes. Commercial growers pick tomatoes before they ripen, then they are shipped. The plant hormone ethylene causes tomatoes to ripen during shipment. An artificial gene can transcribe an antisense RNA complementary to the mRNA for the enzyme used in ethylene production. Consequently, less ethylene is produced to delay the softening of ripe tomatoes.

OUT OF AFRICA

Roughly 3.5-4.5 billion years ago, a molecule had the ability to make copies of its self. This replicator, perhaps RNA, is the predecessor to DNA. As the atmosphere developed between 3.3 and 2.8 billion years ago, oxygen made photosynthesis and respiration possible. Until recently, life forms during this period were a mystery since it is extremely rare to find fossils 3 billion years old. However, fossil DNA is abundant. MIT researchers Lawrence David and Eric Alm estimate 27 percent of all presently existing gene families originated during this period referred to as the Archean Expansion.[20] Genes for electron transport and respiratory pathways acquired through the symbiosis of mitochondria and chloroplasts made plant and animal life as we know it possible.

Darwin built a case for a single evolutionary lineage. Until 1990, taxonomists divided the phylogenetic tree into two groups; prokaryotes which are single celled organisms without nuclei and organelles, and eukaryotes which are multicellular life with nuclei and organelles.

In 1990, Carl Woese, a microbiologist at the University of Illinois, proposed a Three Domain System based on sequences of highly conserved ribosomal DNA (rRNA). Comparing rRNA is especially useful because throughout nature it carries out the same function, and its structure changes very little over time. The new organization of the phylogenetic tree separates prokaryotes into *eubacteria* and *archaea*. Microorganisms now comprise the majority of the planet's biological diversity and now represent two of the three domains of life.

Ribosomal RNA reveals *eubacteria* and *archaea* are distantly related. Scientists estimate prokaryotes appeared roughly 3 billion years ago. *Eubacteria*, include bacteria which have circular chromosomes. *Archaea* are the most abundant organisms in the ocean and were perhaps the earliest life form. Many *Archaea* live in extreme environments surviving extremes of temperature, radiation, pressure, salinity, and acidity.

Eukarya includes unicellular and multicellular plants, animals, fungi, amoeba, and protists. Fungi include yeast and molds which obtain nutrients through absorption not photosynthesis. Protists include slime molds, algae, and protozoans. Scientists estimate eukaryotes appeared roughly 2 billion years ago.

Approximately 580 million years ago, the Cambrian explosion led to the major body plans in animals. From these major body plans, scientists estimate that mammals appeared roughly 200 million years ago and *Homo sapiens roughly* 200,000 years ago. Scientists estimate that humans split with the chimpanzee, *Pan troglodytes*, roughly 6 million years ago. Major human differences include delayed maturation, less body hair and muscle strength, increased cognitive capabilities, and the ability to walk upright and use complex language.

When comparing chimpanzee and human genomes, the major difference is that humans have one less chromosome. At some point two human chromosomes fused at their telomeres through a mutation referred to as a

Robertsonian translocation to make chromosome 2. In addition, the two genomesy have close to the same number of protein coding genes with one-third having identical sequences. At the genome level we differ in 1.2 percent of protein coding genes and that rises to about 4 percent when non-coding DNA is included.[21]

Genes that appear to have evolved more rapidly in humans than chimpanzees are transcription factors. Gene regulation explains how the relatively small number of genetic changes has led to the anatomical and behavioral differences in chimpanzees and humans. One gene variation is the *FOXP2* on chromosome 7 involved in the appearance of speech.

To explain the differences between modern humans, several theories have emerged. First, *Homo sapiens* replaced *Homo neanderthalensis* (Neanderthals) which evolved independently roughly 350,000 years ago in Eurasia and disappeared roughly 30,000 years ago.

Second, modern humans evolved in both Africa and Eurasia with gene flow between populations. Svante Paabo of the Max Planck Institute for Evolutionary Anthropology sequenced extracted DNA from bones of Neanderthals and estimates a 1-4 percent similarity suggesting interbreeding between modern humans and Neanderthals.[22]

Third, In *The Origin of Races*, Carleton Coons proposed distinct human races exist including Caucasoid, Mongoloid, Negroid, Capoid (Southern Africa), and Australoid. Based on Craig Venter's analysis of the diploid genome, the similarity among humans at the genomic level is 99.5 percent, so you will find population characteristics, but race does not exist in the genetic code.

Harvard University population genetecist Richard Lewontin's statistical analysis of genetic data assembled on the diverse populations of the world reveals that most of the detectable genetic variation in the human species is between individuals in the same population, estimated at roughly 85 percent. Another 9 percent is between populations assigned to the same ethnic group, while differences among ethnic groups constitute only about 6 percent of the genetic variation of the human species.[23]

Fourth, fossil evidence suggests our species originated in East Africa. Studying DNA samples gathered from diverse populations around the world has provided strong evidence not only for the Out of Africa theory.

Mitochondrial DNA which is located in the cytoplasm has its genome passed on through the maternal line and does not recombine with paternal DNA. Consequently, mtDNA remains similar to its ancestral DNA with the exception of mutations. These mutations are traced to analyze evolutionary lineages.

In 1987, Allan Wilson and Rebecca Cann, University of California-Berkeley geneticists, and Mark Stoneking of the Max Plank Institute analyzed mtDNA from 147 human DNA samples gathered from around the world for the Human Genome Diversity Project. A computer analysis of mutations revealed that mtDNA samples from *Homo sapiens* are derived from a single female named referred to as Mitochondrial Eve living in East Africa over 200,000 years ago.[24]

Similarly, the Y chromosome in males does not recombine with maternal DNA, so mutations in the Y chromosome are used to trace the lineage of the male line. In search for what Spencer Wells refers to as scientific Adam, a number of studies on the Y chromosome from different parts of the world reveal polymorphisms originate from a single mutation also found in East Africa.[25] Researchers determined that mutations in the Y chromosome occur at a slower rate than in mtDNA. Adjusting for the mutation rate in the Y chromosome, Luca Cavalli-Sforza's lab at Stanford University estimated migrations out of sub-Saharan Africa began roughly 50,000 years ago.

CHAPTER 7

Crossing the Weismann Barrier

In 1809, Jean Baptiste Pierre Lamarck's proposed two laws to explain biological diversity. First, the use and disuse theory for the restructuring of vital forces to adapt to the environment, and second the traits acquired through environmental cues are transmittable to offspring.

In 1887, August Weismann proposed the germ plasm contained in human chromosomes is isolated from somatic cells providing a barrier from environmental influences. Evelyn Fox Keller observed, "When Weismann's concept was incorporated into Mendelian genetics, geneticists succeeded in purging Lamarckism from science."[1] Today, we know plants, fungi, bacteria, unicellular animals, and some multicellular animals do not have separation between somatic and germ cells.[2]

Austrian biologist Paul Kammerer revived Lamarckism in the 1920s. He experimented with midwife toads that mate in the hot and dry terrains of Europe and Northwest Africa. The derivation of their name originates from the act of males taking the strings of eggs laid by females and carrying them on their legs and backs to the water. Kammerer forced the toads to live in water, and subsequently the offspring preferred to live and mate in water. In order to adapt to the new environment, the male toads acquired black nuptial pads to grasp on to females. Kammerer claimed that the nuptial pads are acquired traits inherited from the father although he provided no mechanism for his theory.

It is important to understand Lamarckism in historical context. Lamarck was a vitalist and didn't think in terms of a physical unit as the concept of DNA, genes, and chromosomes were not yet understood. After analyzing the human genome, scientists have revived Lamarckism again. Today, with a better understanding of basic science through epigenetic regulation and cultural inheritance via niche construction, scientists are able to provide evidence for Kammerer's claim.

EPIGENETIC REGULATION

In the 1940's, developmental biologist Conrad Waddington coined the term epigenetics combining genetics and epigenesis to explain how an organism is formed gradually from an embryo. Waddington defined epigenetics as environment-gene interactions that induce developmental phenotypes.

Epigenetic inheritance allows an organism to respond to the environment without having to change its hardware. Randy Jirtle of Duke University uses a computer analogy to understand epigenetics. The genome is the hardware, and epigenetics works similar to computer software.[3] Through epigenetic regulation, genes remain constant; however, they are switched on and off to make differentiated cells. Scientists want to better understand the mechanisms of epigenetic regulation on traits and diseases, gene expression, cell differentiation through programming of stem cells, and human development.

Most human cells contain a copy of the entire genome including all its genes. Egg and sperm cells contain half the genome, and red blood cells and some white blood cells which have no DNA are the exceptions. However, not all genes are switched on or transcribed in all cells.

Stemness refers to what it is that makes the roughly 220 types of cells found in humans different. Epigenetic information plays an important role in turning genes on or off during the cell cycle. Human embryos develop from embryonic stem cells that originate from a blastocyst, a ball of cells that develop into differentiated cells. Epigenetic mechanisms occur naturally so that every gene is not expressed in every cell.

Researchers have discovered three main mechanisms for epigenetics; DNA methylation and genetic imprinting which occur at the DNA level, and histone modifications that occur at the protein level allowing developmental control of gene expression (see Table 7.1).

Table 7.1 Mechanisms for Epigenetics

DNA methylation the covalent addition of a methyl group to cytosine
Histone modifications transcriptional repression by proteins through chromatin remodelling
Genetic imprinting gene expression depends on which parent passed on the gene

DNA METHYLATION

In 1975, Robin Holiday and John Pugh discovered DNA methyltransferases (DMTs), the enzymes that bind methyl groups to cytosine nucleotides.[4] Methyl donors can silence genes via enzymes that bind methyl groups onto DNA. DNA methylation, the covalent attachment of methyl groups, is the most common covalent modification of the human genome.

Scientists estimate 40-45 percent of human DNA is methylated to deactivate foreign DNA including transposons and retrotransposons which maybe harmful.[5] Methylation protects against foreign endonucleases, enzymes that cut DNA at specific points, mainly from bacterial DNA. If a base pair is methylated then the specific restriction enzyme will not recognize the DNA. Since human DNA and bacterial DNA are methylated differently and in a specific manner, this defense mechanism allows ensures only foreign DNA is destroyed by endonucleases.

DNA methylation is important in mammals for reprogramming germ cells and essential during development. DNA methylation can provide stability and also a mechanism for switching genes on or off during development. Cytosine exists in normal and methylated versions of genes. Researchers have discovered heavily methylated genes in cells are not expressed, and cells that have non-methylated forms of these genes are expressed.

Phenotypic plasticity is the ability of an organism with a given genotype to change its phenotype to adapt to its environment. Insect castes differ physically and behaviorally, but not solely based on their genomes. Individual insects contain the instructions needed to develop into each caste, but only when the appropriate genes are activated. Royal jelly secreted from the glands of worker bees determines which bee larvae become queens by silencing particular genes. Queen bees live for several years and are fertile producing many offspring. In contrast, when the same genes are active, they become workers. Workers bees eat pollen and nectar; have stingers, and a lifespan of approximately a month.[6]

Mammals are diploid, thus have two alleles for each gene. Epigenetic modifications can determine which allele is expressed. Robert Waterland of Baylor Medical Center and Jirtle have used mice to demonstrate that certain alleles are particularly susceptible to changes in methylation from maternal nutrition. The researchers fed folic acid and vitamin B12 to pregnant mice to investigate differences in gene expression. Mice given the methyl rich supplements had offspring with mostly brown fur, whereas mice without supplements gave birth to offspring with yellow fur. The methyl groups from the supplements switched off the *agouti* gene which gives mice a yellowish coat. The DNA in the *agouti* gene which harbors a transposable element remained the same; however, the offspring inherited the methyl modification.[7]

In humans, epigenetic changes affecting our gene expression are also determined by what our mothers and grandmothers ate during pregnancy. The amount of folate in human breast milk or commercial formulas determines the expression of alleles of certain genes. Studies in humans have linked normal spinal cord development and cognitive abilities to methylation patterns.

HISTONE MODIFICATIONS

DNA is coiled around proteins called histones which form chromatin and provide sites where additional gene regulation occurs. The mechanism for transcriptional repression is through chromatin remodelling. DNA is less likely to be transcribed when tightly bound by histone proteins. Cells allow access to a stretch of DNA by signaling a histone to open up.

After translation, proteins can undergo post-translational modifications through attachment with other chemicals. These modifications attach functional groups and alter proteins physical and chemical properties. Post-translational modifications through chemical modifications by acetylation, phosphorylation, methylation, and ubiquitination inhibit transcription referred to as a histone code. For example, the histones in heterochromatin (genetically inactive DNA) are generally unacetylated while those in euchromatin (protein coding genes) are acetylated (an acetyl group added) which causes relaxation.

GENETIC IMPRINTING

Genetic imprinting refers to the expression of a gene and a resulting trait depending upon which parent passed on the gene. As a result, if a brilliant male scientist is marries a beautiful female model, the result will not necessarily result in a beautiful, brilliant child. This phenomenon is also important for understanding certain traits and diseases since they do not follow the predicted ratios of Mendelian genetics. For example, the rare diseases Angelman, Prader-Willi, and Beckwith-Weldemann syndromes are caused by either maternal or paternal imprinted genes.

If you inherit a Y chromosome from your father, you are a male, an X chromosome you are a female. Everyone receives an X chromosome from their mother; consequently females have two X chromosomes. In 1961, Mary Lyon proposed the random inactivation of one female X chromosome to explain female coat color in mice.[8] To prevent twice as much expression from X linked genes in females, one X chromosome (either the maternally or paternally derived X) is silenced during the early development of the female embryo. This theory was confirmed when the inactivated X chromosome was visible during interphase of mitosis as a condensed chromosome called a Barr body.

To better understand the mechanism of inactivation, scientists investigated the phenomenon of a callipyge sheep, one which develops large buttocks. Mysteriously, only 10 percent of its offspring possessed the same trait not the expected 50 percent. After mating both genders, Jirtle discovered only sheep that inherit a mutated copy of the callipyge gene from the father and two normal alleles from their mother have large muscular bottoms. After further investigation, Jirtle discovered a polymorphism, a single base mutation from A to G found in junk DNA, was linked to the affected sheep and blocks the expression of a gene.[9]

CULTURAL INHERITANCE

Unlike most plants and animals that have a narrow environmental range in which they can survive, our species has proceeded to inhabit almost every portion of the earth. In response to environmental changes including hotter temperatures and scarce water, roughly 50,000 years ago, our ancestors began migrating out of southeast Africa. These migrations have subsequently led to tremendous genetic diversity.

In order to adapt to diverse environments, humans developed innovative survival techniques. These adaptations coincided with a cultural revolution spurred by greater human intelligence.[10] Jared Diamond refers to this development of our increased intelligence and brain size as The Great Leap Forward.

Genomics researchers have linked positive selection to adaptive lifestyle changes during the Late Pleistocene and Holocene. Approximately 14,000 years ago, global warming occurred at the end of the Pleistocene Ice Age. The Stone, Bronze, and Iron Ages followed the Ice Age are characterized by humans manufacturing tools.

As societies became more complex, humans acquired technologies as part of their culture. In the process it changed our ancestor's diet, how they defended themselves, and their exposure to diseases.

Using tools, organisms can modify their environment creating the selection pressures to which they are exposed. For example, bird's nests and spider's webs are the result of the organism's own niche construction. A group of Oxford University researchers observed, other than humans beavers manipulate their environment considerably more than any other species.

> A beaver's dam modifies many selection pressures in the beaver environment, some of which are likely to affect the fitness of genes that are expressed in quite different traits, such as their teeth, tails, feeding behavior, susceptibility to predation, diseases, and social systems.

> The environment for which the beavers are being selected is in part the product of their own activity. Modifying natural selection pressures in environments gives feedback in the evolutionary process resulting in different evolutionary traits. Niche construction allows acquired characteristics to play a role in the evolutionary process, not in a Lamarckian way, but by their influence on selective environments through feed back to select their own genes.[11]

Craig Venter estimates a 99.5 percent similarity in human DNA based on the diploid genome. Among the 0.5 percent differences are roughly 15 million single nucleotide polymorphisms (SNPs) which are nucleotide variations that make humans unique. The distribution of polymorphisms across populations reflects our human history through natural selection, genetic drift, and mutations.

Evolutionary geneticist Spencer Wells describes our DNA as the greatest of all history books. As part of a five year Human Genographic Project which began in 2005, Wells and other specialists from the fields of linguistics, climatology, anthropology, and archaeology became detectives to provide a more complete picture of our past by attempting to decode that history book.[12] What is coded in DNA is the story of human history, migrations, and evolution.

Positive natural selection is the force that drives the prevalence of alleles that increases fitness, and it has played a central role in our development as a species. The ability to detect natural selection in humans greatly improves the study of human history and medicine.

In the past, scientists compared individual candidate genes for positive natural selection. Today, genome wide sequencing and polymorphism data provides new candidates for selection.[13] Genomics researchers can now compare DNA sequences in populations to determine if the mutations are due to natural selection. Finding uninterrupted DNA segments is strong evidence of recent adaptation. Genes still linked to their neighbors suggests that they are in the gene pool due to natural selection, not by chance. Researchers have linked positive natural selection to cultural variations. A high rate of positive natural selection has occurred in the past 10,000 years because of urbanization and increased population density.

Roughly 10,000-12,000 years ago hunter-gathers began to practice agriculture and domestication of animals. Our ancestors discovered by recombining DNA of food sources by interbreeding two varieties of livestock or plants yielded more desired traits. Plant breeders learned to pick the fittest plants and transfer pollen. After roughly six generations the desired traits become fixed, referred to as hybrid vigor, a phenomenon still not fully understood by scientists. Through food selection, culture has changed our digestion processes and susceptibility to diseases.

Researchers have found evidence of strong selection in the lactase gene coinciding with the domestication of cattle 5,000-10,000 years ago in parts of Africa. After nursing, humans do not naturally produce lactase, the enzyme produced in the small intestines to digest lactose or milk sugar. The gene is programmed to turn off. Cattle domestication led to mutations disabling the molecular switch that turns off the production of lactase. Those with the mutation are able to drink milk their whole life, while others are lactose intolerant.

In a 1986 analysis of the DNA of several cytochrome enzymes in mitochondrial DNA of the gray wolf and domestic dogs, UCLA evolutionary biologist Robert Wayne discovered only a 0.2 percent difference. This confirmed our ancestors domesticated and breed wolves for desired traits. When Asians and Africans domesticated the gray wolf, it was a trade off. They bred it for desired traits to provide services including, protection, killing pests, and pulling heavy objects. In return, the gray wolf received a home, food, and protection. Today, the gray wolf's relatives include wild wolves, wild dogs, and over 300 breeds of dogs which are house pets and companions.

As hunter-gathers acquired agricultural practices and domesticated farm animals 10,000-12,000 years ago, the population density increased. Urbanization provided greater exposure to animal pathogens and an opportunity for viruses to find hosts and wreak havoc upon humans through the spread of infectious diseases. Some diseases occur only in settled populations. It is necessary to have several hundred thousand susceptible people to sustain endemic diseases such as small pox, typhoid, yellow fever, measles and tuberculosis.[14] Recent genetic changes in the human genome include genes that have responded to infectious diseases. Measles and small pox coincided with the domestication of farm animals and agriculture.

For plant cultivation, Africans cleared the rain forest and planted food staples such as yams and maize. The clearings created more standing water increasing the breeding grounds for malaria carrying mosquitoes. Humans exposed to malaria developed resistance which increased the frequency of the allele for sickle cell anemia, an anemic blood disorder. The genes targeted for selection are frequently found in those with African ancestry.[15]

Genetic variations in organisms result from environments, cultures, and the use of technologies chosen by the organism changing the evolutionary process. As a result, the co-evolution of gene expression and ecological niches through changing selection pressures that the descendants are exposed to plays an important evolutionary role.

Using linkage disequilibrium, researchers discovered nucleotide variations are sometimes found in blocks called haplotypes which are inherited together. This finding contradicts Mendel's Law of Independent Assortment which states units of heredity have an equal chance of inheritance. Haplotypes that are long and common are a signs of selection. When selection works on one locus, it will influence gene frequencies at linked loci a phenomenon referred to as hitch-hiking. Researchers have discovered four haplotypes of sickle cell hemoglobin suggesting four separate malaria endemics.

Using 1.6 million SNPs, researchers have found an estimated 1800 genes are the result of recent selection.[16] Using the HapMap, a database of known haplotypes, University of Wisconsin Anthropologist John Hawks estimates that in the past 5,000 years, positive selection has occurred at a rate roughly 100 times higher than any other period of human evolution.[17]

CHAPTER 8

The Genomics Bubble

When James Watson and the NIH sold the public Human Genome Project to Congress in 1988, there was a bargain. Based on estimates from top scientists, to sequence and map the human genome, the government needed approximately $3 billion in taxpayer's money. That equates to $200,000,000 annually for fifteen years, the largest budget of any biological project in history.

In return, the medical research community aimed to mine genomes for gene variants to locate susceptibility factors called biomarkers in hopes of better understanding diseases and ultimately provide drugs, therapies, and vaccines that lead to their treatments. The medical field currently relies predominately on indicators such as antibodies, body temperature, blood pressure, cholesterol, low-density lipoproteins (LDL), and metabolites in blood or urine samples to make decisions regarding treatments.

Ideally, this information would provide the foundation for pharmacogenomics or personalized medicine. Due to genetic differences, some Asians have alcohol intolerance and some individuals suffer side effects from drugs such as penicillin. Similarly, pharmacogenomics researchers aim to detect reliable biomarkers in the form of gene variants and SNPs that provide correlations to a patient's disposition to drug reactions and metabolism.

According to Duke University genomics historian Robert Cook-Deegan there was a second component to the sales pitch.

> Science administrators and members of Congress who shepherded the budgets for genome research (and their counterparts in other nations and international organizations)

supported the project not only because of its medical benefits, but also because they saw it as a vehicle for technological advance and creation of jobs and wealth. The main policy
rationale for genome research was the pursuit of gene maps
as scientific tools to conquer disease, but economic development was an explicit, if subsidiary, goal.[1]

Was the public Human Genome Project successful in meeting either of
these goals of providing medical benefits and a boost to the biotechnology
and pharmaceutical industries justifying the use of taxpayer's money?

In 1997, Juan Enriquez and Rodrigo Martinez introduced the term
bioeconomy which refers to economic activity derived from genetic and
molecular industrial processes. Currently, the pharmaceutical-biotech industry is growing faster than other sectors of the economy. The U.S. GDP
was roughly $14 trillion in 2007 and grew at an estimated 2.2 percent.
Statistics compiled by Rob Carlson indicate the pharmaceutical-biotech
industry accounts for an estimated $250 billion in sales annually in the
United States with a growth rate of 6-8 percent and $600 billion world-
wide.[2]

Carlson's statistics further reveal biologics, medicinal products created by biological processes, is growing even faster. Biologics grew in the
range from 15-20 percent.[3] Biologics accounts for approximately 25 percent of worldwide pharmaceutical-biotech revenues and 30 percent of U.S.
revenues with the remaining revenues derived from agricultural and industrial products.[4]

Currently, the biotechnology industry as a whole is thriving, however,
it is difficult to determine what impact genomics has made on medicine.
Despite this complexity, Battelle, which is an independent organization, released a genomics impact study in 2011. Battelle's study points out that the
projected investment of $3 billion provided by American taxpayers from
1988-2003 was actually $3.8 billion, and from the years 1988-2010 that this
investment has resulted in an estimated $796 billion impact on the Ameri-
can economy.[5] The $796 billion figure includes sectors other than medicine
such as forensics, computing, environmental science, and industrial processes. The $796 billion figure also includes revenues provided by thou-
sands of jobs and millions in taxes from all of these sectors.

While the analysis of technical data is not conclusive for medicine,
an analysis of fundamentals is more revealing. While most scientists expected an acceleration of treatments for diseases, ironically the pharmaceutical industry has experienced a small number of new drugs, therapies

and vaccines in the pipeline. In a statement issued by the Mayo Clinic regarding the current status of pharmacogenomics:

> In the future, pharmacogenomics could have an expand-
> ing role in the practice of medicine. But despite the prom-
> ise of personalized medicine, pharmacogenomic testing is
> not widely available today. You should be skeptical of news
> reports and other sources of information proclaiming that
> pharmacogenomics or other types of personalized medicine
> will offer revolutionary results today. It is hoped that this
> will be true sometime in the future.[6]

This statement does not reflect of the potential of genomic medicine, rather the current status. The current status of genomic medicine raises the question of what obstacles the field currently facing? These obstacles raise the further questions. What roles, if any, have the scientific community and the U.S. government played in creating these obstacles? Also, what strategies are the scientific community and the U.S. government using to address these obstacles?

SCIENTIFIC DOGMA

Although the medical community is struggling with developing medical treatments, scientists have benefited from the discoveries in basic science generated from the Human Genome Project and analyzing variations in human genomes. These discoveries have led to a scientific revolution, a paradigm shift in the understanding of the concept of a gene. Consequently, the approach to understanding the mechanisms of complex diseases and evolution has shifted from genes to a dynamic process.

Although natural historians began studying biological systems centuries ago, it was not until the 20th century that researchers began to fully understand hierarchal levels and the robustness of biological systems. In the early 1900s, Walter Cannon investigated homeostasis in physiological systems. In 1948, MIT mathematician Norbert Wiener combined cybernetics, Greek for steersman (*kybernetes*), with biology. Bio-cybernetics enabled researchers to better understand the linkages and complexities in a biological system by measuring the effects one component has on another. In 1968, Ludwig von Bertallanfy formulated the General Systems Theory based on the dynamics, self regulation, and hierarchal structure of systems.

In the 1960s, the publishing of Rachel Carson's *Silent Spring* sparked an environmental movement. The movement is concerned with the impact humans have on the environment. This led to the National Environmental Policy Act (NEPA) of 1970 to establish requirements for environmental impact statements on projects significantly affecting the quality of the environment. The Environmental Protection Agency or another responsible federal agency must prepare a statement describing the full environmental impact for the public. In the statement, biologists model ecosystems to analyze the potential effects of a proposed project.

According to University of Texas Philosopher of Biology Sahotra Sarkar, in the past scientists did not generally accept cybernetics in molecular biology; however, in the twenty-first century Wiener's vision has returned with a vengeance.[7] Today, systems biologists integrate different types of data rather than reducing it (see Figure 8.1).

To better understand the complexity of biological systems, systems biologists capture and integrate as many hierarchal levels of biological data as possible including parts, their functions, and interactions. Research data ranges from DNA and RNA, protein, protein-protein and protein-DNA interactions, signaling and regulatory networks, cells, organs, individuals, populations, to ecologies.[8] Systems biology provided the impetus to perturb biological systems and develop the tools to quantify reactions at a number of hierarchal levels.

In the last decade, a number of private institutes and university programs have incorporated integrative and systems biology. In 2000, Hiroaki Kitano established the Systems Biology Institute in Tokyo and Leroy Hood established the Institute for Systems Biology in Seattle. Duke, Harvard, MIT, and UC Berkeley have established systems biology departments. In the U.S., the National Science Foundation and the NIH have allotted funding for grants to researchers in systems biology research to support these programs.

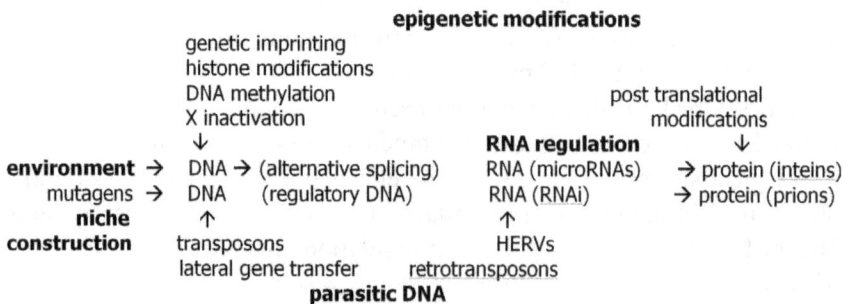

Figure 8.1 Mechanisms of molecular genetics

REGULATORY UNCERTAINTY

In the private sector, researchers require incentives to invest in the discovery of disease related genes. In 1999, these incentives were questioned when the Associated Press and Reuters reported that at a briefing White House Press Secretary Joe Lockhart mistakenly announced that the USPTO had ruled scientists can no longer obtain patents on genes.

Despite the USPTO and White House's attempts at damage control, the pharmaceutical and biotech industry lost tens of billions of dollars in stock value. Celera's stock plummeted 21 percent, Incyte 27 percent, and Human Genome Sciences 19 percent.[9] On March 15, 1999, the NASDAQ stock index dropped 46 percent; its second steepest dive ever resulting in tens of billions of dollars lost market value. When biotech stocks plummeted, investors were temporarily reluctant to invest in the industry.

The erroneous announcement was, however, foretelling as President Bill Clinton and Prime Minister Tony Blair would advocate and later announce open access to the human genome on a NIH online database diminishing the profitability of gene patents. Although Craig Venter was able to sequence the human genome more efficiently than the public project, Celera's business model failed miserably since pharmaceutical companies using genes as drug targets were not paying for gene sequences.[10] Since the business model for private companies is based on incentives provided through intellectual property, Venter subsequently moved on from Celera which became a drug discovery and development company.

One person who was willing to invest their time and money on disease related genes is former Harvard researcher Kari Stefansson. In 1996, the scientist turned businessmen founded deCODE Genetics which is based in Iceland. Iceland provides an ideal research infrastructure for genomics research since the Icelandic government keeps a database of medical records with the family history of roughly 260,000 Icelanders.

deCODE's business model is based on identifying genes associated with common diseases using population studies to develop drugs. deCODE has received funding from joint ventures with Merck and Applied Biosystems. Roche Pharmaceuticals guaranteed $70 million to deCODE with an additional $130,000 in payments resulting from agreed upon milestones. Roche also granted free drugs to Icelanders for any drugs resulting from the research. deCODE has also received a number of patents on genes and processes, but no milestone payments were paid.[11] Rather, the company went bankrupt and restructured in 2009.

Among deCODE's accomplishments is deCODEme, the first personal genomics web-based service with more than one million nucleotides used for direct to the consumer (DTC) genetic testing. The DNA chips manufactured by companies such as Illumina and used by genomics companies such as Celera and deCODE were originally intended for research purposes. With the declining costs of sequencing genomes, this service has provided a potential a new avenue for private sector profits.

The Food, Drug, and Cosmetic Act (1938) authorizes the FDA to provide regulatory oversight for the safety and efficacy of biological products and medical devices. The services that personal genomics companies provide are classified as a medical device. Consequently, personal genomics companies are required to have a permit to sell genetic testing services directly to the public.

The government recognizes individuals should have access to their genomes; however, is still working through several regulatory issues related to DTC genetic testing. The government's concerns include analytical validity or accreditation of the medical devices, scientific validity or the access to advice from qualified professionals to interpret the data, and the method of informed consent given by the consumers.[12] Personal genomics companies obtain informed consent from consumers through websites or the mail, which is legal. The FDA's concern is that this type of informed consent bypasses ethics committees and IRB approval.[13]

With other types of genetic testing such as prenatal amniocentesis, paternal testing, and neural tube defects; consumers have access to genetic counseling or advice from a physician. With genomic screening for diseases, the medical community currently lacks the knowledge to adequately provide advice. Although biomarkers for some diseases including Alzheimer's and macular degeneration are good indicators for high risk, most diseases lack reliable indicators. Consequently, genetic testing kits screen for raw data and do not provide information related to reactions to medications or provide medical claims of clinical utility.

In response to these concerns, the FDA sent letters to five personal genomics companies — 23andme, Navigenics, deCODE Genetics, Knome, and Illumina with a warning that they are manufacturing and selling medical devices without appropriate premarket review and approval. At the state level, the New York and California State Health Departments have banned the direct sale of genetic tests to consumers until companies acquire licenses.

Although some personalized genomics companies are profitable, an uncertain future for DTC genetic testing will exist until formal regulatory guidelines are adopted. Pathway Genomics' personal genomics division pathwayandme had planned to sell its genome testing kits which utilize saliva samples through 6,000 of Walgreen's 7,500 store drugstore chain.

However, the lack of a stable and coherent regulatory structure has caused Pathway Genomics' to postpone its plans.

With the soaring costs for drug development, the FDA has recognized the need for more interagency cooperation as well as collaboration between the FDA, researchers, and pharmaceutical companies to break down barriers between regulators and industry. Despite the U.S. government's attempts to remove barriers to providing efficient medical treatments, the controversies over intellectual property and incentives to the private sector and policies regulating devices used in personalized medicine have created a climate of regulatory uncertainty in the pharmacogenomics industry.

BURSTING THE BUBBLE

To address the genomics bubble, in 2004 the Food and Drug Administration issued the document *Innovation or Stagnation: Challenge and Opportunity on the Critical Path to New Medical Products.*[14] The FDA's strategy is to modernize medicine by taking advantage of advances in science and technology, bioinformatics, and imaging. In addition, the FDA seeks to address any administrative shortcomings that affect the ability to utilize these advances in the drug approval process.

In the document, the FDA outlines the Critical Path Initiative for modernizing medicine. The strategy is to improve the product development process through critical path, basic, and translational research. In 2006, a FDA Task Force convened to implement the strategy for critical path research improving the product development process for new medical treatments.[15] This plan addresses product safety and effectiveness through developing tools for assessing safety standards.

In order to study gene-disease relationships, biomarkers are essential as clinical tools in the development of safe drugs. There are three ways in which pharmacogenomics is applied in clinical practice; identifying responders of treatment, determining appropriate dosages of responders, and by identifyng susceptibility to adverse drug reactions.[16] In addition to measuring individual therapeutic responses, biomarkers also predict disease progression and assist in evaluating disease processes. Creating better standards for biomarker validation will improve the clinical trials process.

In 2008, the President's Council of Advisors on Science and Technology recommended longitudinal population studies to effectively correlate genomic and clinical data. DNA samples in databases are representative of populations, but not necessarily the disease. Researchers can utilize DNA specimens over time to study the progression of a disease, comparing those with and without a disease, and comparing populations for risk factors to validate whether or not genetic risk factors have clinical utility.[17]

In 2007, then Senator Barack Obama of Illinois sponsored the Genomics and Personalized Medicine Act paving the way for biobanks, the repositories for biological specimens collected for medical research. This bill authorized and provided appropriations for NIH to establish, collect, and maintain the biorepositories.

The National Human Genome Research Institute organized the Electronic Medical Records and Genomics Network (eMERGE), a national consortium, with a research strategy to combine DNA biorepositories with electronic medical records.[18] Using data from the expanding computerized electronic medical records network over a lifetime and tissue samples from biobanks, researchers can mine patient data and compare outcomes.

In addition to understanding gene-disease relationships, in 2004, Francis Collins expressed the need in the US for a similar research effort to better understand gene-environment interactions.[19] Researchers can accomplish the task utilizing case-control studies of people including representative populations with and without a particular disease monitored over time.

Another aspect of biomarker validity is determining the detection rate. By developing a database of drug reactions and polymorphisms for screening, researchers hope to minimize the adverse reactions and deaths. A study published in the *Journal of the American Medical Association* reveals that roughly 106,000 people die each year in American hospitals as the result of side effects from medication.[20]

BASIC TO TRANSLATIONAL RESEARCH

The FDA has recognized that medical product development has not kept up with advances in basic science.[21] This gap has stimulated a greater emphasis on translational research which takes basic scientific discoveries from the lab to the clinic. With fewer genes than expected found in the human genome and a better understanding of disease processes, researchers are focusing on other mechanisms that contribute to diseases. Among those are disease pathways; RNAi, microRNA, epigenetic, telomerase, and stem cell therapies; and DNA vaccines.

In addition to pharmaceutical companies and university medical research centers working with specific diseases, several initiatives were created to pursue translational research on a larger scale. The Translational

Genomics Research Institute (TGen) in Phoenix, Arizona is a non-profit organization established in 2002 set up to take advantage of the advances derived from genomics. TGen currently has drugs, therapies, and vaccines for specific diseases in clinical trials.

NIH has plans to establish its own center for translational medicine in 2012 contingent upon Congressional funding of a $730 million budget. NIH's proposed National Center for Advancing Translational Sciences will place the existing related government programs under one roof. According to a NIH press release, the intentions of the federal drug development center are not to compete with private industry, but rather to collaborate.[22]

CHEMICAL PATHWAYS AS DRUG TARGETS

Curing diseases by identifying genes and searching for drug targets is more complex than originally thought. Relying solely on genetic differences among individuals has not provided a wealth of information regarding individuals and drug reactions. For this aspect of personalized medicine the key is to select treatments based on changes in metabolic processes and signaling pathways.

Most drug targets fail in animal testing and only 10 percent in the pipeline make it through clinical trials.[23] Blockbuster drugs are expensive to produce costing nearly a billion dollars and require 10-15 years for FDA approval. Pharmaceutical researchers use various methods to develop drugs. Steve Hall, who heads research and development at Serenex, Inc. in Durham, North Carolina uses an innovative target based screening process called proteome mining. It is a serendipitous process of finding targets which can be screened with automated high throughput equipment. A common process used by other pharmaceutical companies is cell based screening which utilizes living cells to develop a lead candidate for clinical trials.

An alternative to using SNPs for drug development is to identify pathways as potential drug targets. Pathways consist of chemical reactions that work together to control cell functions including metabolism, regulation and signal transduction, and the cell cycle. Most of the targets for new drugs are proteins which have a unique 3D structure that can interfere with disease pathways. Many cellular proteins are molecular switches called enzymes. The principle switches are kinases. These enzymes have receptor sites that bind to molecules called ligands. Ligands are prime targets for drugs to manipulate cellular function by inactivating enzymes that pathological microbes and cancer cells need to live.

Research on medical disorders reveals common pathways among

some diseases. While at Genentech Susan Desmond-Hellmann developed Herceptin, which targets aggressive breast tumors formed by genetic mutations that upregulates the gene *HER2*. A type of stomach cancer also has the *HER2* mutation and is also responsive to Herceptin. A type of brain cancer and a type of skin cancer share the hedgehog signaling pathway which means that the same therapy may provide treatments for both diseases.[24]

Understanding the mechanisms that regulate body weight and possible drug therapies to treat obesity are of great interest to pharmaceutical companies. In 1994, Dr. Jeffery Friedman and a team of Rockefeller University researchers identified the first recessive mouse obesity (*ob*) gene which travels in cells through the bloodstream to the brain signaling fullness. A mutation in the gene makes mice become fat. Humans have an almost identical gene which produces a hormone called leptin (derived from *leptos*, the Greek word for thin) which regulates fat storage.

Greater levels of leptin are found in individuals with more fat and reduced levels in those who diet. However, not all obese patients have increased levels of leptin which suggest there may be important differences in the cause of obesity.[25] Some people have a mutation in the receptor which prevents leptin from binding to suppress appetite.

Friedman had a theory that some obese people produced leptin at a greater rate to compensate for a faulty signaling process. When researchers injected leptin into diabetic mice and they did not lose weight, the researchers confirmed that the mice had defective leptin receptors.[26] Similarly, when researchers at Amgen investigated potential applications of leptin for weight control treatment, they discovered that some obese people have high levels of leptin in their blood because their leptin receptors are desensitized.

Research on behavioral disorders also reveals common pathways among diseases. Researchers at Columbia University discovered common regulation of energy metabolism and bone mass. Serotonin is a hormone that controls appetite. When the leptin-serotonin pathway is turned on in mice, researchers found the appetite increases resulting in weight gain and increased bone mass. When the pathway is turned off, mice eat less and lose weight, and their bones weaken providing a link to the same pathway for osteoporosis and obesity.[27]

While researching mental disorders, scientists have found no simple link with genes. Through linkage analysis, what researchers have found is a statistically significant correlation of genetic inheritance in some diseases including Alzheimer's, schizophrenia; bipolar disorder, addiction, and depression.

Approximately 19 million American adults are diagnosed with depression each year. Researchers discovered childhood adversity, alcohol dependency and substance abuse play a contributing role. A polymorphism in the promoter region of the serotonin transporter *5-HTT* gene is found to moderate stress. Individuals with one or two copies of a short allele exhibit more symptoms of depression in relation to those homozygous for the long allele.[28]

Using brain cells from rats and mice, scientists use knockouts, deleted genes in a process of elimination, to search for susceptibility genes. Scientists have linked the *5-HTT* gene that codes for the protein that escorts serotonin across synapses, the spaces between brain cells, to anxiety in monkeys and humans. Serotonin nerve pathways are involved in controlling emotions, sleep, anxiety, and aggression. Prozac and Zoloft, drugs created to treat depression, act on the serotonin receptors.

Drugs alter behavior and physiology through different mechanisms. Heroin addiction has susceptible polymorphisms found in the opioid receptor gene. Nicotine and cocaine's addiction primary pathway is dopamine. Also, the desensitization of dopamine receptors and alterations in synaptic transmission are adaptations to counteract the altered brain function from alcohol addiction. Research reveals alcoholism and eating disorders also have overlapping pathways.

DNA VACCINES

Since humans are unable to fight off many infectious diseases naturally, immunologists developed effective vaccines for infectious diseases such as the measles, polio, tetanus, mumps, and smallpox. The immune responses with conventional vaccines are not produced by the host, but are injections of antibodies formed by other animals. Surface proteins are used to identify targets for antigens. A sheep or mouse is injected with an antigen to produce large quantities of monoclonal and polyclonal antibodies which are extracted. Traditional vaccines use a weakened form of a pathogen so the host's immune system, specifically B cells will produce antibodies against the antigen. Over time, as the vaccines immunity decreases immunity, boosters are needed.

Today, people around the world are burdened by a number of different types of infectious diseases without cures. Scientists are currently unable to produce efficient vaccines against malaria, HIV, herpes, hepatitis B, AIDS, SARS, malaria, and replicating strains of bird influenzas. This is because some pathogens reside inside cells. Viruses are intracellular parasites. Traditional vaccines have strong antibody immunity, but weak cellular immunity.

Analysis of gene sequences of pathogens can reveal microbial surface proteins that are potential antigens for use in developing DNA vaccines. In contrast to conventional vaccines, DNA vaccines also provide cellular immunity. Immunologists are attempting to inject DNA coding for an antigen and its promoter into the host's nucleus, where it is transcribed into the cells DNA which produce pathogenic proteins. One hurdle to developing DNA vaccines is the host's immune system sees it as a foreign antigen. Another hurdle is that DNA vaccines are limited to the protein components of pathogens. Some pathogens including those that cause pneumonia, staff infections, and meningitis have a protective outer shell made of polymerized sugars, called polysaccharides.

Table 8.1 Comparison of conventional and DNA vaccines

Conventional Vaccines	DNA Vaccines
antibody immunity (B cells)	antibody (B cells) and cell immunity (T cells)
antibodies from animals	host produces antibody
use surface proteins	use genome of pathogen
limited life, need boosters	do not require refrigeration and ability to have multiple diseases inoculated in a single vaccination

Researchers are currently testing DNA vaccines in animals. The first approved DNA vaccine is for the West Nile Virus in horses. The first cases of West Nile Virus were reported in Uganda in 1937. The first cases in the United States were reported on house pets and farm animals in 1999. In 2005, the U.S. Centers for Disease Control and Prevention reported 2,949 cases and 116 West Nile Virus related deaths in the United States. Scientists believe the virus which can cause encephalitis or meningitis is caused by mosquitoes that first bite an infected bird migratory bird then bites a human.

RNAi and MicroRNA Therapies

The discovery of RNA interference (RNAi) which is a natural defense mechanism against parasitic viruses and transposons also has potential applications for antiviral therapy. Some cells treat small interfering RNA (siRNA), roughly 20 single stranded nucleotides, as a virus infection. The cell uses siRNA as a guide to find and destroy single stranded messenger RNA that have the same sequence. This coding mechanism prevents viruses from using the cells genetic code to reproduce.

When a specific gene is linked to a disease, injecting siRNA can prevent the mRNA from coding for a protein. The siRNA will bind to the mRNA produced by that gene. These synthesized nucleic acids are termed antisense because its base sequence is complementary to the gene's messenger RNA, which is referred to as the sense sequence. The sense strand, the strand with the same sequence as a target gene, is removed leaving the antisense strand to silence the gene. When the strands bind, this prevents the ribosome from gaining access to mRNA and the production of protein.

One advantage of RNAi over conventional drugs is that antisense molecules are extremely specific. Conventional drugs bind directly with disease causing proteins, but their non-specificity may lead them to bind with other proteins, resulting in side effects.

Despite the potential of RNAi therapy and spending hundreds of millions of dollars on its development, pharmaceutical companies Roche, Pfizer, and Abbott have decided to scale back the development of RNAi drugs in favor of alternative therapies. The biggest challenge is delivery. RNA is quickly broken down in the bloodstream and has trouble entering cells. To survive in the bloodstream and avoid immune responses, researchers are chemically modifying RNA.[29]

Studies have also linked microRNAs, post-transcriptional regulators that silence genes, to potential therapeutic roles in cancers, heart disease, diabetes, and neurological disorders. In 2011, researchers at ETH Zurich discovered two microRNA molecules that affect insulin signaling which lead to obesity and type 2 diabetes.[30] When the researchers used a viral vector to express the two microRNAs in healthy mice, they developed high blood sugar. The researchers also found that silencing the two microRNAs in obese mice improved glucose sensitivity. This discovery has the potential to provide obesity treatments in humans.

EPIGENETIC THERAPY

Scientists have linked epigenetic mechanisms to cognitive disorders, obesity, diabetes, heart disease, and cancer. Consequently, pharmaceutical companies are interested in reprogramming our genes as medical interventions. Researchers with the Human Epigenome Project are building a database of epigenetic modifications resulting from gene-environment interactions to better understand complex diseases. The project's goals are to better understand methylation patterns and histone modifications, and utimately catalog the number of human genes that are turned on and off by viruses, diet, and environmental toxicants such as smoke, exhaust, pesticides, and heavy metals.

Researchers have discovered a delicate balance exists between DNA methylation and cancer. The increase in the methylation of promoter regions of tumor suppressor genes has resulted in an increase in the number of tumors.[31] Hypermethylation of C-G islands (successive cytosine-guanine dinucleotides) prevents transcription factors from binding and leads to the inactivation of tumor suppressor genes, which include the *BRCA* genes linked to breast cancer. The role of tumor suppressor genes is to maintain DNA in cells. Usually repairs are made before the cell is allowed to divide. Without these repairs, cells with damaged DNA will copy the damage and pass it on as a permanent mutation in future generations of new cells. The deactivation of a tumor suppressor gene reduces the cell's ability to check for damage at DNA check points.

Too little methylation is also harmful. In 1983, researchers at Johns Hopkins discovered the loss of methylation at C-G dinucleotides in tumor samples. Insertions of viral oncogenes into host DNA causes mutations in tumor suppression genes and the deletion of DNA. If an oncogene loses the ability to regulate the cell cycle, the cell may divide uncontrollably. Normal cells have proto-oncogenes which are activated from too little methylation.

STEM CELL THERAPY

Individuals with cystic fibrosis lack of protein that regulates chloride ion transport through a cell membrane resulting in the accumulation of mucus in the lungs. Huntington's is a neurodegenerative disease linked to a mutation that produces tri-nucleotide repeats. Genetic diseases such as cystic fibrosis and Huntington's are candidates for gene therapy. However, gene therapy has had limited success in clinical trials due to immune response, problems with viral vectors, and gene regulation.

An alternative therapy for genetic and degenerative diseases is replacement cells. While the potential for stem cell therapies is great, the moral and legal barriers have rivaled the scientific hurdles. The last three U.S. Presidents have faced lobbies from both sides of the stem cell debate. In 1994, President Bill Clinton simply placed a moratorium on embryonic stem cell research until further discussions took place.

President George W. Bush appointed primarily communitarians, which have neoconservative religious affiliations to his eighteen-member Council on Bioethics.[32] As expected, the members voted overwhelmingly against unrestricted embryonic stem cell research and the cloning of embryonic stem cells. Bush was, however, the first President to legalize embryonic stem cell research.

Bush's 2001 stem cell research policy prohibited the use of federal funding for embryos destroyed after August, 9, 2001. However, his policy allowed federal funding for research and cloning of existing embryonic stem cell lines and obtaining stem cells by other methods including umbilical cords, animals, and adults. Although the decision mobilized activists on moral grounds, to minimize legal issues, Bush's policy required that donors must provide informed consent and not benefit financially.

In 2009, shortly after becoming the President, Barack Obama issued an Executive Order ending Bush's restrictions on federally funded embryonic stem cell research. This resulted in more grants proposals to NIH using donated embryos from fertility treatment centers. In a turn of events, on August 23, 2010, U.S. District Judge Royce Lambeth issued an injunction banning NIH from funding human embryonic stem cell research.

In September, the U.S. Appeals Court temporarily waived the injunction until oral arguments were heard. After oral arguments in September, three judges ruled that federal funding on human embryonic stem cell research is allowed while the appeals process continues. The major issue the judges must decide is if human embryonic stem cell research violates the Dickey-Wicker Amendment (1996) which prohibits federal funding if human embryos are destroyed.

Lambeth's ruling would make the prior use of destroyed embryos illegal and place current stem cell research in limbo. Specifically, the U.S. government has approved two trials on humans using embryonic stem cells; Advanced Cell Technology for macular degeneration and Geron Corporation for treating spinal cord injuries. In April 2011, a US Federal Court of Appeals in Washington ruled in favor of allowing federal money for funding human embryonic stem cell research, overturning the August 2010 injunction.

TELOMERASE THERAPY

In 1880, Walter Flemming discovered mitosis, the process that allows our somatic cells to duplicate and grow. Chromosomes break during the mitosis. While studying maize chromosomes and discovering transposons in the 1940s, it became apparent to Barbara McClintock that chromosome ends must have a mechanism to prevent them from fusing together.

In 1988, Robert Moysis at Los Alamos National Laboratory sequenced human telomeres, the repeating nucleotide sequences (TTAGGG) at the ends of chromosomes that can reach a length of up to 15,000 base pairs. Each time a non-cancerous somatic cell divides during mitosis, some of the telomere is lost, usually from 25-200 base pairs. The telomeres on the ends of each chromosome are controlled by telomerase. Telomerase, a specialized reverse transcriptase enzyme, elongates chromosomes by adding (TTAGGG) sequences allowing a cell to continue to grow and divide. When telomeres become too short, chromosomes can no longer replicate and somatic cells die.

Telomerase is is used as a diagnostic tool for identifying cancers. Telomerase detected in cancer cells is significantly more active than in normal body cells. Without telomerase, telomere shortening eventually prevents the growth of cells. Cancer cells inhibit telomere shortening which enable uncontrollable cell division resulting in a tumor. The lack of telomerase activity provides a mechanism for tumor suppression. Researchers want to discover a way to safely turn off telomerase in humans. Geron Corp. currently has breast and pancreatic cancer drugs that inhibit telomerase in FDA clinical trials.

Researchers have also discovered a correlation between telomere length and life span. Mice live roughly three years and their somatic cells divide from 14-28 times during their lifetime. The Galapagos turtle's somatic cells divide from 90-120 times and live over 150 years. In 1961, Leornard Hayflick grew human somatic cells in the laboratory and noticed they have the capacity to reproduce a finite number of times, roughly fifty rounds of cell division referred to as the Hayflick limit. After comparing telomeres in different types of cells, Moysis discovered older people have shorter telomeres. Cells from humans in the 80-90 age group can live in culture for roughly 20 divisions.[33]

Ronald DePinho of Harvard Medical School and the Dana-Farber Cancer Institute genetically altered mice to lack telomerase which prevents telomere shortening. The mice aged prematurely and exhibited numerous health related problems. Restoring telomerase reversed the aging process and none of the mice developed cancer following the treatment.

Spanish researchers increased telomere expression and telomere length in mice through the telomerase pathway using the supplement TA-65. In female mice, the researchers were able to reverse cellular aging and improved health indicators including glucose tolerance and osteoporosis without a significant increase in cancer.[34]

In 1997, researchers at Geron Corporation isolated a gene that encodes the protein/enzyme telomerase. Using telomerase therapy researchers at Geron Corporation and Texas Southwest Medical Center were able to make two types of human cells reproduce past their Hayflick limits.[35] Telomerase therapy entails extracting adult stem cells are and inserting telomerase then returning the modified cells back into genes. With this procedure, scientists can potentially reverse the aging process and extend human life.

CHAPTER 9

What Can Economists Learn from Evolutionary Biology?

A housing bubble peaked in 2005 which set the stage for the 2008 financial crisis. The financial crisis had a ripple effect on the housing, travel, retail, automobile, and oil industries and led to increased unemployment which contributed to the 2008 recession.

A role of economists is to provide an economic analysis of recessions. Both Keynesian economists, who advocate government intervention, and market economists, who advocate free markets, agree that the effects of globalization especially in Asia, outsourcing, trade deficits, and unfavorable currency exchange rates are partially responsible for the current recession. However, Keynesians and market economists have different perspectives to the basic questions of what caused the 2008 recession.

Economists also propose strategies for restoring economic prosperity. With automobile manufacturers and financial institutions on the brink of collapsing, a plunging stock market, and a freeze in the credit markets, the federal government reacted with bailouts and a stimulus. In contrast to Keynesian economists, market economists argue that in free market/ capitalist societies, economic growth and wealth creation are better fueled by entrepreneurship and technological innovation than through a stimulus. In this regard, industrial revolutions are the best strategy for restoring economic prosperity.

In 1996, Paul Krugman, who received the 2008 Nobel Prize for Economics, delivered a talk in Europe titled *What Economists Can Learn from Evolutionary Theorists*. Although a major critic of the non-equilibrium model, according to Krugman, he is a self-described student of evolutionary economics. Evolutionary economics compares the parallels between evolutionary and economic models and provides an interesting analogy to compare the Keynesian and market approaches to economics. Evolutionary economists view the economy in non-equilibrium with circular and cumulative causation similar to a dynamic ecosystem. In contrast, Keynesian or neoclassical equilibrium economics is compared to classical physics which is in equilibrium.

A MARKET OR GOVERNMENT FAILURE?

Free markets came under attack because of the 2008 recession. Keynesians blame Wall Street and the real estate industry for their greed while market economists blame Keynesian macroeconomic policies and the lack of regulatory oversight ultimately led to the recession. Consequently, no consensus exists among economists on whether market or government failures occurred. An analysis of the events leading up to the housing bubble and 2008 financial crisis reveal both accusations have merit.

THE HOUSING BUBBLE

Economic bubbles are characterized by high volumes of business at prices that are considerably higher than intrinsic values. Americans experienced the dot-com bubble when the market crashed in 2000 due to overconfidence in technology. Investors took their money out of the stock market and purchased real estate which was a better investment opportunity. Subsequently, many people purchased homes they could not afford and others have lost their jobs due to the recession.

Following the Great Depression, Franklin Roosevelt signed the Glass Steagall Banking Act of 1933. This Act provided banking reforms including the FDIC and the separation of Wall Street and consumer banking. With the repeal of Glass Steagall in 1999, banks were able to expand operations which contributed to an increase in sub-prime loans.

In *Meltdown*, Thomas Woods Jr. proposes macroeconomic policies first initiated during the Carter Administration are the origins of the housing bubble.[1] The Community Investment Act lowered qualifying standards for homebuyers. In addition, Keynesian economists including 2008 Nobel Laureate Paul Krugman suggested more affordable financing to further increase home ownership. Subsequently, the Federal Reserve lowered interest rates on several occasions. Consequently, leading up to the 2008 financial crisis, government intervention led to inflation and over building, and economic indicators became unrealistic.

With his ability to manipulate lawmakers, James Johnson, Chief Executive of Fannie Mae, was the architect of the private-public home ownership program.[2] Under this program, borrowers received loans many originating from Countrywide Financial and Nova Star Financial which were backed and processed through the federal programs Fannie Mae and Freddie Mac. This program's loans were processed with unrealistically low qualifying standards which led to a high percentage defaulting. Subsequently, the American taxpayers were responsible for the bill, a government failure.

THE FINANCIAL CRISIS

In the early 1990s, Bankers Trust invented credit default swaps, a type of over-the-counter derivative, as a way for banks to unload the default risk of their asset portfolios while also lowering capital requirements.[3] On several occasions, it became apparent that these financial instruments were potentially destructive. In 1994, Orange County, California public officials speculated on interest rate swaps, a type of over-the-counter derivative, purchased from Merrill Lynch. The investors lost almost $2 billion gambling with taxpayer's money.

From 1995-1997, a Connecticut based hedge fund Long-Term Capital Management (LTCM) had 46 percent, 40 percent, and 20 percent returns.[4] In 1998, LTCM began losing hundreds of millions of dollars on a daily basis. These losses were fueled by Russia defaulting on bonds.[5] Rather than intelligently diversifying their investments, some Wall Street banks invested too heavily in credit default swaps accumulating toxic assets which led to significant financial losses, a market failure. Without regulation on capital reserve requirements for compliance with stress tests, LTCM was able to leverage $5 billion into more than a trillion dollars.[6]

While Chairman of the Commodity Futures Trading Commission (CFTC) from 1996-1999, Brooksley Born was concerned about the lack of transparency in the multi-trillion dollar over-the-counter derivatives markets. Born was interested in regulating the derivatives markets; however, she faced several hurdles. The financial lobby had five lobbyists for each congressman and the large banks had given millions in campaign contributions.[7]

Perhaps the major roadblock Born faced was the President's Working Group on Financial Markets. The Working Group included CFTC Chairman Born, Federal Reserve Chairman Alan Greenspan, Treasury Secretary Robert Rubin who was formerly Chairman of Goldman Sachs, Deputy Treasury Secretary Larry Summers, and Chairman of the Securities &

Exchange Commission (SEC) Arthur Levitt. Unlike Born, when testifying before Congress all the others promoted deregulation and were successful. As a result, the SEC, CFTC, and state regulators currently lacked jurisdiction over the OTC derivatives markets.[8]

As is with the case with President Bush's appointed conservative Council on Bioethics, Roy Cardato of the John Locke Foundation points out, market-based and free market public policies are not the same. Policy makers can manipulate the market for their desired outcome, and when this happens a government failure occurs.[9]

For markets to work smoothly there is a need for regulation due to externalities such as fraud that affect the whole financial system as well as individuals. Authorities are currently investigating possible collusion between banks and credit rating agencies, and brokers selling commodities knowing that they would tank.

By 2008, the derivatives markets would reach $680 trillion.[10] In an October 23, 2008 Congressional Hearing discussing the financial crisis, Rep. Henry Waxman grilled former Federal Reserve Chairman Greenspan on the. Greenspan admitted flaws exist in his philosophy of free markets regulating themselves, specifically resisting the regulation of derivatives. Greenspan had argued the economy was as strong as ever and regulations would interfere with the economic prosperity.[11]

Rather than having a planned economy or a decentralized *laissez faire* economy, the optimum economic system is obtained through managed markets. Congress has drafted legislation in an attempt to address the regulatory gaps in the financial industry, a government failure, for transparency and requirements for leveraging capital. Congress is also investigating past fraud and conflict of interest with traders and rating agencies, a market failure, and drafting proactive legislation to prevent further fraud and conflict of interest. To address these issues, President Obama signed the Dodd-Frank Wall Street Reform and Consumer Protection Act in 2010. The Wall Street Transparency and Accountability Act of 2010 is in the process of Congressional approval and the President's signature.

A MARKET OR GOVERNMENT SOLUTION?

Keynesian and market economists also disagree on which approach to take in order to revive the economy. Curiously, the Nobel Prize Committee has awarded prizes to economists with theories supporting contradictory approaches for understanding and managing national economies. Guy Sorman, an adjunct scholar with the Manhattan Institute, provides a unique perspective to why this enigmatic phenomenon has occurred.

Sorman's *Economics Does Not Lie* begins, "Economics is a science, whose purpose is to distinguish between good and bad policies."[12] But is economics actually a science? Traditionally, economics is thought of as a social science. If economics has become more of a science with the aid of quantitative tools such algorithms, mathematical models, and computers to interpret data, economists should have the ability to analyze both approaches to determine which system works best and provide guidance during recessions.

In *Architects of Ruin*, Peter Schweizer points out that many of the Washington, DC power players that lobbied for social engineering policies during the Clinton Administration are now dealing with the current financial crisis and recession.[13] Keynesians prescribe a government intervention to resurrect the economy. President Obama and Congress approved a $787 billion stimulus package via the American Recovery and Reinvestment Act of 2009.

With casualties including Bear Stearns, Merrill Lynch, Lehman Brothers, American International Group (AIG), General Motors, Ford, and others the federal government committed $700 billion in Troubled Asset Relief Program (TARP) funds, roughly $250 billions in stock equity purchases to the major investment banks with the largest amount $40 billion going to American International Group (AIG) and $20 billion each to Citigroup and Bank of America, and $82 billion in loans to bail out auto makers. A government stimulus may stimulate the economy in the short run, but not necessarily in the long term due to the staggering debt.

CONFLATIONARY EVOLUTIONARY ECONOBABBLE

Equilibrium and perfect competition are the foundation of Keynesian or Neoclassical economics. Keynesian economics is based on the assumption that individuals are fully informed to make rational choices to maximize utility and firms have the ability to maximize profits. Friedrich Hayek critiqued the Keynesian model of economics because it is based on utopian conditions. Hayek cast doubts on a centralized states ability to set pricing, since perfect knowledge of investor's desires is required. Even with computers, it is virtually impossible to manage an unlimited amount of data which is not even available to the state.[14]

Krugman refers to evolutionary biology and economics as sister fields because they have a remarkable amount in common; the questions they ask, the methods they use, and the way they relate to the world.[15] Krugman's talk begins:

As you probably know, I am not exactly an evolutionary economist. I like to think that I am more open-minded about alternative approaches to economics than most, but I am basically a maximization-and-equilibrium kind of guy. Indeed, I am quite fanatical about defending the relevance of standard economic models in many situations. Why, then, am I here? Well, partly because my research work has taken me to some of the edges of the neoclassical paradigm.

But there is another reason I am here. I am an economist, but I am also what we might call an evolution groupie. That is, I spend a great deal of time reading what evolutionary biologists write - not only the more popular volumes but the textbooks and, most recently, some of the professional articles.

My interest in evolution is partly a recreation; but it is also true that I find in evolutionary biology a useful vantage point from which to view my own specialty in a new perspective. In a way, the point is that both the parallels and the differences between economics and evolutionary biology help me at least to understand what I am doing when I do economics — to get, to be pompous about it, a new perspective on the epistemology of the two fields.

The Keynesian model is based on equilibrium similar to the physical sciences. In contrast, a group of market economists referred to as evolutionary economists advocate an economic model that parallels evolutionary biology which is in non-equilibrium. This model exhibits circular and cumulative causation, and also evolves.

Evolutionary economists want to get away from the idea that individuals maximize and the notion of equilibrium. In particular, they want to have an approach in which things are always in disequilibrium, in which the economy is always evolving.

Latterly there have also been some economists who want to

merge evolutionary ideas with the Schumpeterian notion that the economy proceeds via waves of creative destruction. And if you are a reader of Gould and his acolytes, you have the sense that evolution proceeds through spasms of sudden change that seem positively Schumpeterian in their drama.

According to Krugman, most evolutionary biologists model the natural world not as being on the way, but as being already there. He further says:

> The most telling example of this preference is the widespread use of John Maynard Smith's concept of "evolutionarily stable strategies." An ESS is the best strategy for an organism to follow given the strategies that all others are following - the strategy that maximizes fitness given that everyone else is maximizing fitness, with each taking the others' strategies into account. Does this sound familiar? It should, the concept of an ESS is virtually indistinguishable from an economist's concept of equilibrium.

> Now you can understand why I say that a textbook in evolution reads so much like a textbook in microeconomics. At a deep level, they share the same method: explain behavior in terms of an equilibrium among maximizing individuals.

The following year, Krugman launched another attack against evolutionary economics in *Slate*. This time it was specifically aimed at a 1997 Cato Institute conference on bionomics. Bionomics is an economic discipline which studies the economy as a self organized ecosystem. Krugman accuses Michael Rothchild of the Bionomics Institute and author of *Bionomics: Economy as an Ecosystem* of not understanding evolution or economics and refused an invitation to the conference.[16]

THE ADAPTATIONIST PROGRAM

Ironically, some of the strongest attacks against Darwinian evolution are from within the biology community. According to evolutionary biologist Lynn Margulis, neo-Darwinism, which insists on the slow accrual of mutations, is in a complete funk. The scientific community accepted evolution in the 1940s; however, an adaptationist program dominated evolutionary thought in England and the United States coinciding with the Modern Synthesis. The Modern Synthesis provided a mechanism for Darwinian natural selection to occur through mutations and Mendelian inheritance. The Modern Synthesis assumed microevolution, the change of allele frequencies from selection, mutation, genetic drift, and migration are the only modes of evolution.

Two Harvard professors, paleontologist Steven Gould and evolutionary geneticist Richard Lewontin, critiqued the adaptationist program in the classic paper *The Spandrels of San Marco and the Panglossian Paradigm* (1979) using hypothetical relationships between adaptations and selection in mice.[17] Gould and Lewontin's attack of the adaptationist paradigm was a reaction to two books published in 1975, E.O. Wilson's *Sociobiology* and Richard Dawkins' *The Selfish Gene*.[18]

In the paper, Gould and Lewontin articulated a number of reasons why we should consider alternatives to adaptationism or natural selection as the sole source of evolution.

NO SELECTION AND NO ADAPTATION
The gene for white fur may also protect the mice from a virus. If the virus wipes out the mice with black fur, the mice with white fur have a selective advantage. However, the selective advantage is not because the fur is white. The trait may result from genetic drift or may result from no selection and no adaptation.

SELECTION WITHOUT ADAPTATION
A genetic mutation that increases fecundity may affect natural selection, but is not necessarily adaptive. With fewer natural resources, the individuals will leave no more offspring, but twice as many eggs with the excess dying because of resource limitation.

ADAPTATION WITHOUT SELECTION
Although a trait is adaptive, it is not necessarily a product of selection. For example, the color of mice's fur may result from diet rather than from genetics. The white mice can adapt to a white sandy environment, but not due to selective forces.

In the Spandrels article, Gould and Lewontin examine the adaptationist hypothesis using the spaces between the archways supporting the domed roof of the basilica of St. Marks Cathedral in Venice. The archways were not designed as artistic objects, but as necessary architectural by-products of mounting a dome on rounded arches.

In *Candide*, Voltaire ridiculed Dr. Pangloss, saying things can not be other than they are. They ask, "Were our noses made to carry spectacles and our legs clearly intended for breeches?[19] The point — you cannot always correctly assume that the current use of a trait is the reason that trait is present. "The spandrels became exaptations (non-adaptive spaces) utilized for a function other than that for which it was developed through natural selection. In this case, the old features are used for new purposes, the canvasses for the residency of mosaic designs."[20]

PUNCTUATED EQUILIBRIUM

Darwinian evolution accounts for mutations in a population's gene pool that are gradual which occur over many generations. While microevolution is a change in allele frequency, macroevolution is a rapid change in a genome at the species level. In contrast to adaptationists, evolutionists argue that evolution arises from a number of interacting processes not solely through the interaction of an organism and the environment leading to random mutations and natural selection. In 1972, based on Ernst Mayr's 1954 paper on geographic speciation, Harvard paleontologist Niles Eldredge and Gould argued evolution can occur rapidly which they referred to as punctuated equilibrium.[21]

A rapid reorganization of genes can result through a major climate change from global warming or asteroids. Following a dramatic climate change, the introduction of new species into an ecosystem leads to filling numerous ecological niches referred to as adaptive radiation. Paleontologists discovered in the fossil record that rapid speciation and the development of major body plans for animals occurred during the Cambrian explosion roughly 580 million years ago. Along with EvoDevo, this phenomenon partially explains the lack of gradualism in the fossil record.

Another source of acquiring DNA and a mechanism for speciation is through symbiosis. By comparing genomes, researchers are better able to understood organelles in plants and animals. In plants, chloroplasts have separate DNA from the plants nuclear DNA. Chloroplasts contain chlorophyll that use light energy to form carbohydrates from CO_2 and water during photosynthesis. Scientists believe that *Cyanobacteria*, a blue-green bacterium, with ribosomal protein genes which are closely related to chloroplasts, are its precursor.[22]

Mitochondria, the power stations in animals that combine food and oxygen to form ATPs, are circular DNA located in the cytoplasm of a cell. Scientists compared the genomes of the bacteria *R. prowazekii* and mitochondria using a ribosomal protein gene. The comparison revealed an identical set of genes for the synthesis of ATP. Maternally inherited mitochondrial DNA consists of over 16,000 base pairs and contains thirty-seven genes.[23] The finding indicates the bacterial genome entered the cytoplasm of an ancient eukaryotic cell as the result a symbiotic relationship.

Today, scientists understand evolution as the interaction of many processes with gradual and rapid changes at a number of levels. Macroevolution can occur as a result of the cumulative effects of microevolution, and can happen in a single generation. In Krugman's analysis of evolutionary economics, he conflates micro and macro evolution. While microevolution follows the equilibrium model with successful adaptations, macroevolution follows the non-equilibrium model with rapid changes in genomes.

INVEST AND INVENT

With a full plate of pressing public policy issues to contend with including energy, global warming, healthcare, the recession, and national security; polling reveals the economy is the major concern of most citizens. This is predominately due to failed macroeconomic policies which have created the need for a new paradigm for economic growth.[24] Analysis of recessions such as the one created by the current housing bubble and financial crisis requires the government to choose whether to back a failed infrastructure or adapt a different strategy by investing in a new one.

In market economies, innovation and destruction are equally important. In the 1930s, economist Joseph Schumpeter coined creative destruction describing the phenomenon of markets acting as a selection vehicle eliminating obsolete technologies and utilizing innovative ones.[25] Businesses that do not adapt successful practices and employ creative people do not survive. When comparing economics to evolutionary biology, punctuated equilibrium or the rapid changes in genomes are analogous to creative destruction.

Technology can incrementally overhaul industries, for example, the news industry moving from print to web or the music industry evolving from records, eight tracks, cassettes, and CDs to iPods. The transition can also involve a radical transformation. At the turn of the twentieth century, the transportation industry employed hundreds of thousands of carriage makers and blacksmiths. Subsequently, automobiles replaced horses and now the automobile industry is transitioning to electric cars and biofuels. In addition to transportation, creative destruction has also revolutionized weapons, communications, health care, medicine, and agriculture.

With industrial revolutions, some individuals are worse off due to layoffs and bankruptcies. In the free market model, when technological change creates economic disruptions, economic indicators are not necessarily predictive of how the economy is performing, rather just points in the business cycle. In this scenario, government intervention is not necessary since markets are self correcting.

Technologies require development and face obstacles in the process. Economists, however, do not agree on the appropriate role for government involvement in innovation and growth. Despite America's current economic problems, our land of opportunity still provides a standard of living envied throughout the world. The best environment for the commercialization of innovation is found in the United States.[26]

Maintaining an efficient infrastructure and the development of technologies are necessary for America to stay competitive at the international level. This process begins with basic research. The NIH and the National Science Foundation were instrumental in providing funds for basic research for genomics and nanotechnology and currently for applied research.

As part of the stimulus package, the American Recovery and Reinvestment Act of 2009 allocates funds for infrastructure, tax incentives, unemployment benefits, healthcare, and education. Among the infrastructure expenditures are public goods including highways, defense, and police protection, and funding targeted towards innovation in medical research and technology. This funding includes $27.2 billion on research for more efficient energy systems, and $7.6 billion on scientific research. As stated in Obama's Strategy for American Innovation, this is necessary and justifiable because:

> The recent crisis illustrates that the free market itself does
> not promote the long term benefit to society, and that certain
> fundamental investments and regulations are necessary
> to promote the social good. This is particularly true in the
> case of investments for research and development, where
> knowledge spillovers and other externalities ensure that the
> private sector will under-invest especially in the most basic
> of research.[27]

> The government has the role as an innovation facilitator
> because basic research does not have direct commercial pay-
> offs. Because basic science has little if any immediate com-
> mercial return, its costs are typically not easily undertaken
> by private investors, thus leaving government funding as a

critical course of support.[28]

After basic research is completed, the government's additional role is to support applied research through private sector innovation. Over the past half century basic and applied science are areas where the United States has positioned itself as the global leader. Using cultural economics, Sorman delves into many layers of analysis to investigate why other developed countries including China, South Korea, India, Japan, Russia, and those in Western Europe lag behind the United States. In addition to the market system, Sorman also credits the quality of institutions (universities, research centers, government, and private), investors and entrepreneurs, and their partnerships for America's phenomenal economic success.[29] So, the development of technologies is best accomplished through strategic private-public partnerships.

Biotechnology is defined as a new era in drug discovery which began in 1976 with the use of genetic engineering distinguishing it from when researchers used various other techniques. According to the Biotechnology Industry Association, the number of biotechnology companies in the United States grew to 1,452 in 2006. This is primarily due to state and regional biotechnology hubs located near universities and venture capitalists.

In the 1970s, financing from the venture capital firm Kleiner & Perkins based near San Francisco and Stanford University was important in a number of the West Coast startups including Cetus and Genentech. This cluster is now referred to as Silicon Valley. Then in 1978, Walter Gilbert, Philip Sharp, and Charles Weissmann started Biogen which introduced the synthetic form of naturally occurring interferon to fight cancer. With the number of research universities in Boston, many more collaborations took place and the second biotech cluster formed.

In 1981, the North Carolina Biotechnology Center, the first state sponsored initiative to develop biotechnology, opened in Research Triangle Park, North Carolina. Its mission is to strengthen academic and industrial research institutions, educate the public, and develop partnerships to move research to commercialization.

Drawing on the successful model of biotechnology hubs, Obama's Strategy for American Innovation provides $50 million in matching grants to support the creation of regional innovation clusters for transfer technology support and to promote collaboration among industry leaders.[30]

So, generally speaking, the government's support for applied research should include investment in the infrastructure and technologies that improve efficiency rather than subsidies. Otherwise, for example, with alternative energy, when the subsidies are withdrawn from this sector, it will not have the capability of competing in the market. If the bubble bursts on a multi-trillion dollar green economy, another economic meltdown could occur.

Obama's Strategy for American Innovation supports open capital markets that allocate resources to the most promising ideas.[31] Among the technologies selected as drivers of future growth are genomics, bio-nanotechnology, and synthetic biology. These biotechnologies are attractive because they can provide job and wealth creation, increase the quality of our lives, and transform multiple industries. They have the potential to improve health care, and create the next generation of plastics, agricultural products, bioremediation organisms, bioweapons, efficient carbon neutral biofuels that will reduce our dependence on oil, novel enzymes, and vaccines.

CHAPTER 10

The Free-Rider Problem

Malaria infects an estimated 300-500 million people annually, and it is fatal to roughly 1.5 million of those. The deadly disease has flu like symptoms; vomiting, headaches, diarrhea, chills, fever, and can cause anemia. Children under five and pregnant women in Africa are the most vulnerable.

Malaria's etymology has its roots in vitalism as *mala aria* is Italian for bad air. In 1880, French Army doctor Charles Leveran observed parasites in the red blood cells of malaria patients. The *Plasmodium* parasites that infect humans are transmitted by female *Anopheles* mosquitoes which feed at night and multiply within the host's red blood cells. The parasite travels to the liver where it is shielded and replicates. After entering into the bloodstream, the parasite invades red blood cells to feed on iron rich hemoglobin. Infected blood cells are destroyed in the spleen; however, the parasites have adapted the ability to display surface proteins on the surface of red blood cells, causing the red blood cells to stick to the walls of small blood vessels blocking the circulatory system.

To fight malaria, initially health care organizations targeted the mosquito's habitat. Mosquito bed nets, head coverings, drainage of marshy areas where mosquitoes lay their eggs, and spraying their habitat with the insecticide DDT, which is now banned, were the main strategies.

Although the medical community has developed a number of drugs, malaria remains a major world health problem. The first effective drug treatment for malaria was quinine which is from the bark of the cinchona tree in the Andes of South America. In the 1940s, Robert Woodward of Harvard synthesized a form of quinine called Chloroquine. The parasite developed resistance to quinine through a mutation resulting in a change in one amino acid. In some regions, sulfur drugs replaced quinine; however, the parasite developed resistance in even a shorter time.

During the Cultural Revolution in the 1960s, the Chinese government launched a project to investigate the properties of plants used in traditional herbal medicines. Chinese herbalists discovered *Qing hao* also known as *Artemisia annua* or sweet wormwood, a plant indigenous to China and Vietnam, is effective for treating fevers.

As a malaria treatment, sweet wormwood's active ingredient artemisinin destroys the *Plasmodium* parasite by releasing oxygen based free radicals into the host's red blood cells. However, soaking its dried leaves and extracting artemisinin is an expensive and laborious process. Currently, the supply is not meeting the demand. Speculators are stockpiling the wild plant, and some reports estimate up to 50 percent of the drug sold in Africa and Asia is counterfeit.[1]

Although *Artemisia annua* has a nearly 100 percent efficacy rate, parasites could develop resistance to one drug. In 2004, the World Health Organization endorsed artemisinin combination therapy (ACT), the use of several anti-malarial drugs which is a more expensive treatment. Scientists devised this approach in order to make it more difficult for parasites to acquire resistance. To develop resistance, it would require that parasites acquire simultaneous mutations.[2]

COLLECTIVE ACTION

Most patients in need of anti-malarial drugs are from developing countries and unable to afford them. Pharmaceutical companies are not anxious to develop treatments for parasitic diseases such as malaria due to the lack of economic incentives. According to Clifford Winston, an economist with the Brookings Institution:

> Market failure occurs when a socially desirable service (that is, one whose social benefits exceed social costs) is not privately offered because it is unprofitable. Market failure also occurs when a service is undersupplied because it is a public good and susceptible to the free-rider problem.[3]

Without incentives to drug manufacturers and investors, and paying customers, there is a need for collective action. In response, outside the box thinker Victoria Hale created the non-profit Institute for OneWorld Health. The former FDA analyst noticed that only 10 percent of the agency's budget went to diseases of the developing world which account for

90 percent of the infections. The non-profit already has had success with three other drugs including paromomycin, an antibiotic for black fever.

With backing from the Bill and Melinda Gates Foundation through a $42.6 million grant, Hale negotiated with patent holders, drug manufacturers, and scientists to form an innovative partnership to treat malaria. Hale's goal is to develop a novel and less expensive process to manufacture the drug. As of 2006, the artemisinin treatment is $2.40 per dose, and the aim is to reduce that to $0.25 per dose. If successful, this will provide an affordable treatment, have an impact on black market drugs, and save millions of lives.

In the 1970s while deciphering the genetic code, H.G. Khorana successfully synthesized the first functional gene using chemical synthesis. The gene has 77 nucleotides and codes for tyrosine transfer RNA.[4] In contrast, University of California at Berkeley professor Jay Keasling, who is working with OneWorld Health, has attempted to develop a novel treatment using synthetic biology. Similar to an engineer redesigning a computer chip to create a different product, Keasling's approach is to create designer organisms by redesigning existing biological systems.[5]

Monad and Jacob of the Pasteur Institute discovered the first gene circuit, a set of genes that help *E. coli* digest lactose.[6] Gene cassettes use feedback loops with bistable toggle switches and feedforward loops with a series of transcription factors regulating each other using inverters made from biological components. Keasling's goal is to insert the genetic code for genes that produce the precursor of artemisinin in bacteria cells by redesigning genetic circuits and metabolic pathways using synthetic DNA. These bioengineered bacteria with synthetic gene circuits will serve as factories used to produce the precursor of artemisinin. The major hurdle Keasling's team faced during the bioengineering process is balancing metabolism.[7] Due to the feedback and feedforward mechanisms in cellular signaling, creating the optimum levels of gene expression requires bioengineering the appropriate regulatory mechanisms.

When the process is refined, Keasling's start-up company Amyris will make the product available to drug makers at cost. Sanofi Aventis is building a bioreactor, a vessel which provides favorable conditions and provides a higher rate of success, to produce the drug by growing gene cassettes in fermenting vats in a process similar to brewing beer. The Institute for OneWorld Health will take the drug through the clinical trials process and FDA approval.

INCENTIVES WITHOUT BARRIERS

Using similar bioengineering processes, researchers also hope to produce microbes to invade and destroy cancer cells, biofuels, and revolutionize vaccine production. If the synthetic biology industrial revolution is successful, it could potentially create a trillion dollar bioeconomy. However, another major challenge is the development of a system that provides incentives to biotech companies that require a source of income or a return to investors while not creating barriers to the development of products for the public good.

Currently, the incentives for researchers vary based on their objectives resulting in different approaches to intellectual property. In the production of artemisinin, the University of California at Berkeley is only interested in treating malaria and has waived its royalties on the vaccine. Keasling's company Amyris can use the principles learned from the process to make profits from future products.

Craig Venter's Synthetic Genomics has partnered with ExxonMobil which provides $600 million to produce low carbon synthetic biofuels using algae and synthetic organisms to sequester carbon dioxide. This for-profit partnership will benefit society by providing a solution to energy independence and reduce carbon emissions which are accelerating global warming.

Some synthetic biology researchers are placing their novel creations public databases. Stanford bioengineer Drew Endy created the BioBricks Foundation to place the oligonucleotides from the International Genetically Engineered Machines (iGEM) compettion in a Registry of Standard Biological Parts created by Tom Knight of MIT which uses Open Source licensing. Endy took this approach because synthetic DNA products are typically made of a number of parts that may each have intellectual property. Although biotech start-ups rely on income through patents, too many patents on individual parts can make the final product cost prohibitive to produce.

Zinc fingers are a common transcription factor family in humans. They bind to a set of three bases, called a codon, which determines if a specific gene is expressed or inhibited. Engineered proteins called zinc finger nucleases bind to a specific gene, breaking both DNA strands, and stimulate recombination to repair a mutation in a gene. Some researchers are concerned that Sangamo's numerous patents zinc finger proteins may hinder progress in developing medical treatments. In response, these researchers have developed the Zinc Finger Consortium creating a public database to circumvent Sangamo's patents and for the scientific community to use.

An option to avoid blocking patents is patent pools. Patent pools are a contractual arrangement agreed on by patent holders to license patents for products such as sewing machines, radios, aircraft, and automobiles that are made up of numerous patented parts.[8] The underlying principle is cooperation between patent holders accelerates product development. Patent pools reduce the costs of innovation and provide an efficient method of obtaining numerous licenses at once through a streamlined process.[9]

In theory patent pools sound attractive; however, James Boyle and Arti Rai of Duke Law School are pessimistic about their use for synthetic biology. With the existing problems of the enabling technologies, computers and biotechnology, they have concluded that the way the United States government has handled software and biotechnology; they will come together as a perfect storm.[10]

In addition to infrastructural problems, patent pools are vulnerable to manipulation. During World War I, patent holders in the aviation industry were reluctant to license their products which slowed the innovation of new products. In 1975, several major holders of aviation patent established a patent pool, colluded to exclude competition, and fixed prices forcing the federal government to intervene and dismantle the arrangement.[11]

ENABLING TECHNOLOGIES

A crucial step to ensuring the success of synthetic biology is the development of its enabling technologies. Most of the enabling biotechnologies for synthetic biology were available in the early 1980s. The discovery of restriction enzymes and reverse transcription were crucial for the development of recombinant DNA technology which made it possible to convert RNA to DNA. Leroy Hood and Frederick Sanger developed prototype DNA sequencers in the late 1970s and early 1980s. In 1983, Kary Mullis invented polymerase chain reaction (PCR) to copy DNA, and Applied Biosystems developed the first DNA synthesizer.

Synthetic biology also relies on fast, powerful, and affordable computing systems to accommodate and analyze extremely large volumes of data. Moore's Law reveals that computing power progresses and its costs decline at a predictable rate. Robert Carlson calculated that the capabilities of DNA sequencers and synthesizers have followed a similar growth pattern to computers referred to as the Carlson Curve. The costs of read-

Table 10.1 Enabling biotechnologies

1980s	Recombinant DNA technology
1970s-1980s	Hood, Gilbert and Sanger invented prototype DNA sequencers
1983	Applied Biosystems developed first DNA synthesizer
1983	Kary Mullis invents polymerase chain reaction

ing and writing new genes and genomes are falling by a factor of two every eighteen to twenty-four months, and productivity in reading and writing is independently doubling at a similar rate. The Carlson Curve projects scientists are approaching the ability to sequence a human genome for $1000 which should occur in approximately 2020.[12]

While the competition during the Human Genome Project created the need for more efficient DNA sequencers, the need for vaccines has accelerated the development of more efficient DNA synthesizers. Based on a known genome, scientists can sequence a virus to better understand its virulence then synthesize it to develop vaccines. With the current status of enabling technologies, we are now approaching the inflection point on the growth curve where scientific breakthroughs in synthetic biology will take place.

Researchers from the University of New York at Stony Brook led by Eckard Wimmer aimed to determine the difficulty in synthesizing one of the simplest known viruses, polio. In 2002, after three years the researchers were able to create the first synthetic virus based on its published 7,500 base genetic code.[13]

In 2003, the Venter Institute synthesized the bacteriophage phi-X174, a virus that infects bacteria, in just two weeks.[14] The researchers used overlapping oligonucleotides with forty bases to assemble its 5,386 base pairs. After injecting the synthetic viral genome into bacteria, it read the instructions and created the virus.

The Spanish Influenza Virus of 1918 created the worst pandemic in history. In 2005, using viral samples recovered from lung tissue of victims in the Alaskan permafrost, scientists at the Centers for Disease Control and Prevention were able to synthesize the virus and its eight RNA genes using reverse genetics creating a recombinant virus by placing it in bacteria.[15]

In October 2010, the J. Craig Venter Institute issued a press release announcing the formation of the Synthetic Genomics Vaccine Institute (SGVI) which has a three year contract with Novartis to produce a bank of synthetically constructed influenza seed strains used for producing vaccines.[16] After identifying a viral strain and its genetic code, researchers are now able to use rapid computerized sequencing and grow large quantities of vaccines in culture. As compared to the current method of using chicken eggs, this method provides a more effective response to seasonal and pandemic flu outbreaks shortening the production time from months to days.

BIOLOGICAL COMPLEXITY

Bioengineers hope to mass produce drugs and as well as other products; however, technologies require development and ultimately a proof of concept. Industrial processes such as synthetic biology face engineering challenges in the process. A major challenge to engineering biological systems is their complexity. Drew Endy's *Foundations for Engineering Biology* elaborates on three engineering principles that scientists can apply to synthetic biological systems.[17]

DECOUPLING

To simplify the complexity of systems, engineers separate complicated problems into simpler problems which are worked on independently. Whether redesigning an existing biological system or designing and fabricating a new system, design and assembly are specialized processes. These specialized processes require decisions at different levels, so it is more efficient to make them independent from each other and not worry about details in other areas of complexity. Decoupling or separating the design and fabrication processes allows scientists to build more complex systems.

ABSTRACTION

In developing biological systems, bioengineers work at increased levels of complexity. They have a continuum from oligonucleotides which are used to build parts, which make devices, which make a system. Bioengineers have abstraction hierarchies and work at one level of complexity without regard for the details of the other.

Table 10.2 Levels of Complexity[18]

Systems
Devices
Parts
DNA

Parts consist of items such as DNA, RNA, protein, and promoters. In combination, parts make devices which have a defined function that is more complex. Parts abstraction makes it easier to model parts which are used in combination with other parts to make devices. Bioengineers aim to create systems level circuitry by developing functional devices that are used in combination with other devices.

Devices include inverters, switches, oscillators, and logic formulas that are used to regulate gene expression, protein function, metabolism, and cell-cell communication. Among the devices bioengineers have successfully developed are; Michael Elowitz and Stanislas Leibler created transcriptional repressor systems regulators[19], M. T. Chen and Ron Weiss created artificial cell to cell communication systems in signaling pathways in yeast which govern basic cellular activities[20], and Maung Win and Christina Smolke created ribozyme switches that control gene regulatory systems and ultimately gene expression[21].

STANDARDIZATION

Similar to the infrastructure of the microelectronics industry where engineers fabricate microchips for circuit boards, bioengineers hope to develop interchangeable biological parts. This requires the identification and standardization of parts such as promoters, ribosome binding sites, and transcriptional repressors.

Drew Endy started the International Genetically Engineered Machines (iGEM) competition coordinated by MIT for undergraduates from around the world to make novel bioengineered products. The students create Bio-Bricks which are 50-100 base pairs of synthetic DNA called oligonucle-otides. Currently, over 5000 BioBricks are available in a public registry, the Registry of Standard Biological Parts. Using the BioBricks, researchers can assemble specialized parts, devices, and ultimately systems.

Bioengineers hope to build synthetic biological systems from compatible standardized parts that exchange information and behave predictably. Endy uses a Lego metaphor to describe synthetic biological parts; however, when parts are assembled, a problem with crosstalk or noise between signaling pathways occurs. Researchers at Harvard Medical School are using multiple approaches to insulating synthetic pathways through modifications and also through the deletion of other reactions.[22]

In *Biology is Technology*, Robert Carlson illustrates a major challenge to designing biological systems using the aeronautical engineering metaphor of geese and early aircraft which appear too heavy to leave the ground. While the solution to the weight and performance issue is resolved by evolution in geese, humans designed efficient systems for aircraft. When aeronautical engineers developed early aircraft, they used a flight model simulation for the interaction of aircraft and the environment. Carlson notes there is nothing similar to a flight model simulation in synthetic biology. This will require the ability to quantify relationships between variables in a model describing phenomenon at the molecular level such as protein bonding, and chemical and physical reaction rates of molecular components.[23]

Through experimentation, researchers will gain a better understanding of basic science, specifically how cells work. Today, the effects of environmental and developmental interaction on gene expression and cell cycles are not fully understood. Until researchers better understood these undesired interactions that nature has already worked out, applications such as gene therapy will have unwanted side effects.

Table 10.3 Foundational milestones in synthetic biology

2000	Elowitz and Leibler's network of transcriptional regulators
2002	Wimmer synthesized polio virus (7,500 bases)
2003	Venter Institute synthesized bacteriophage phi-X174 (5,386 bases)
2005	Chen and Weiss' artificial cell to cell communication
2005	USCDC synthesized the Spanish influenza virus (13,588 bases)
2007	Win and Smolke developed ribozyme switches

A PROOF OF CONCEPT

In 2010, the Venter Institute announced the creation of a synthetic organism. In perhaps innocent coverage by some journalists of the race to create synthetic life created a controversy in terms of what the experiment actually accomplished. For example, the *Guardian* published an article in 2007 titled *"I am creating artificial life, declares US gene pioneer."*[24] By using creating artificial life rather than creating synthetic life in their coverage, their word choice conveys a different scientific meaning.

Venter has had to defend journalist's statements, although technically the resulting criticisms should have targeted misinformed journalists. Artificial describes a bottom-up approach starting with non-living components such as Biobricks and creating a living system. Synthetic or a top-down approach refers to stripping out unnecessary parts or creating a minimal genome and inserting novel genes. In the Venter Institute's article published in the journal *Science* which documents the steps leading to the first synthetic organism, it clearly states a new line of cells were generated which are capable of self-replication, not artificial life.[25]

More accurate coverage by *Scientific American* reveals that the accomplishment of creating the first synthetic organism resulted from stages of experimentation which took place for over a decade and cost over $40 million.[26] In order to achieve their goal, the Venter Institute decoupled the processes of design, synthesis, assembly, and the transplantation of a synthetic chromosome. Venter disclosed that 99 percent of his labs experiments failed before creating a synthetic organism. Venter's lab encountered delays with single base errors, a frameshift mutation, a non-functioning essential gene, and methylation patterns.

According to Daniel Gibson, the lead scientist in developing the syn-

thetic genome, the journey began in 1995 when the Venter Institute sequenced the genome of Mycoplasma genitalium, a bacterium linked to
urinary tract infections.[27] This genome has the smallest number of genes
that researchers can grow in the laboratory. The Venter Institute grew the
bacteria in culture with modifications making it non-infectious.[28]

This minimal genome provides a chassis for building synthetic organisms. Its genome consists of 482 genes comprising 582,970 base pairs, arranged on a circular chromosome. They removed approximately 100 genes
that are dispensable when disrupted and were left with a minimal set of
genes that can sustain life.[29]

The Venter team synthesized a modified version of the minimal 382-
gene chromosome. Using 101 DNA sequences from 5,000 to 7,000 bases
long ordered primarily from Blue Heron Technology, they joined these
pieces together to make larger pieces. They placed the sequences in yeast
and the fragments were assembled by homologous recombination, a process cells naturally use to repair chromosome damage.

Next, the researchers transplanted the synthetic genome into a *M.
genitalium* cell whose own genome has been destroyed to create *M. laboratorium*. However, when attempting to transplant the synthetic bacterial
genome out of yeast and back into bacteria, all the experiments failed.
Bacterial cells use DNA methylation to protect their genome from deg-
radation by restriction enzymes.[30] After the researchers methylated the
DNA, the resulting *M. laboratorium* bacterium was able to replicate itself
with its man-made DNA.

In contrast to journalists and rival scientists downplaying the impor-
tance of the project's scientific and technological advances, the Venter Institute achieved a number of advances that removed barriers to progress
leading to synthetic organisms with beneficial applications (see Table 10.4).
These include the rapid engineering of bacterial chromosomes, advances
in genome assembly, manufacturing larger pieces of functional synthetic
DNA, and improving methods for extracting chromosomes from yeast.

Table 10.4 Milestones achieved in route to *M. laboratorium*

- Created a minimal genome or chassis
- Created first synthetic chromosome (582,970 bases)
- Synthesized a sequence with over a million bases in the lab
- First synthesis of a eukaryote genome
- Created first synthetic organism (1,080,000 bases)
- Successfully transplanted a genome into another species
- Provided a proof of concept for producing cells based upon genome sequences designed in a computer
- Created first synthetic species

CHAPTER 11

The Boo Scenarios

Synthetic biology has the potential to provide "extreme benefits" to society as well as a tremendous boost to the biotechnology industry.[1] However, applications of biotechnologies require public acceptance, the perception that the economic and quality of life benefits outweigh the environmental and public health risks. Historically, humans have had a contentious relationship with technology, some love it and others hate or fear it.

Over the past several decades, bio-Luddites ranging from Ted Kaczynski, FBI codename UNABOMBER (derived from UNiversity and Airline BOMBER), to activist Jeremy Rifkin have attempted to slow down the development of biotechnology. Kaczynski, of the Earth Liberation Front, used illegal means to protest including targeting professors, businessmen, and a geneticist with home-made bombs. Activists such as Rifkin, on the other hand, have decided to fight technology by legal means including scare tactics and lawsuits earning him Time Magazine's most hated man in science.

Failed regulatory oversight of the environment has left legacy issues. In 1962, the release of Rachel Carson's *Silent Spring* would spark an environmental movement. She popularized the claim that in the 1950s-60s DDT, a spray to kill mosquitoes that spread deadly diseases, was also thinning egg shells of birds, and killing bees, earthworms, and wildlife. The EPA subsequently banned DDT in 1972, a decision that remains controversial. The environmental movement was instrumental in developing legislation providing protection from toxins through regulations on pollution. The National Environmental Policy Act (NEPA) of 1970 requires environmental impact statements on significant projects.

In the late 1970s, a chemical leak in Love Canal in Niagara Falls, New York caused a public health emergency. The cleanup involved tearing down two hundred and thirty-seven homes and a school, and the relocation of the families at a cost of $129 million to Hooker Chemical Co. Congress authorized the Superfund program in 1980.

Superfund was supposedly a quick solution to clean up other waste sites and paid for mainly by the polluters. Ironically, over forty of the sites on the original Superfund list were government owned sites. In contrast, the Superfund program has cost taxpayers billions in lawyer's fees and administrative costs, and many sites remain. The public is left with brownfields which are a byproduct of Superfund's flawed liability system. President Clinton called the program not only a government failure, but a disaster.

Another source of the public's negativity towards biotechnology developed through science fiction. In Michael Crichton's novel *Prey* and the movie *Soylent Green*, nanotechnology is used as a means of self-destruction referred to as the goo scenarios. In these sci-fi stories, out of control self-replicating nano-robots are designed and deliberately released to reduce the human population.

Legacy issues and sci-fi scenarios have led to activism. Activists, led by ETC Group advocate the precautionary principle which delays any potential social benefits. Because of activists, those in the biotechnology field fear the overregulation of synthetic biology. Although bioengineer Robert Carlson doesn't downplay the need for risk assessment, he refers to activist's scare tactics as saying "boo."[2]

The federal government has the unique role of promoting science and technology, and simultaneously providing risk assessment of new technologies. Accordingly, public officials are given the task of determining how and at what rate cutting edge biotechnologies will proceed. In May 2000, the J. Craig Venter Institute announced the creation of the first synthetic organism. The Venter Institute's announcement prompted President Obama to request the Presidential Commission for the Study of Bioethical Issues to carefully consider the implications of the synthetic biology field as its first order of business.

In the 1970s, bio-Luddites and activists tried to prevent the use of recombinant DNA technology based on fears of the unknown. Today, synthetic biology is in the similar stage of early development as recombinant DNA technology was several decades ago. At the 1975 Asilomar Conference, scientists took proactive steps to ensure the new field of genetic engineering moved forward safely. What can we learn from Asilomar for analyzing the risks of genetically engineered recombinant DNA that regulators can apply to the risk assessment of synthetic biology?

LESSONS FROM ASILOMAR

Paul Berg's successful cloning of an animal virus was the predecessor to genetic engineering. For genetic engineering, viruses are used to insert RNA into a plant or animal and then transcribed into DNA. Viruses can alter human cells in culture and transform cell lines into a cancerous state. At the time, scientists were concerned with the potential danger of spreading a virus in labs or through the human population if they were misused. Following the introduction of genetically engineered recombinant DNA, California Institute of Technology President David Baltimore declared the field has moved too quickly. He warned, "Since scientific procedures usually take time to perfect, it is wise to be cautious."

At the June 1973 Gordon Conference in New Hampshire, scientists discussed issues related to concerns with genetic engineering, recombinant DNA, and lab safety. Scientists expressed their safety concerns in a letter to the National Academy of Sciences and the National Institute of Medicine which appeared in the journal *Science*. In response, scientists established a panel which replied in a letter published in *Science*, referred to as the Berg letter, with safety recommendations. The Berg Committee called for an international moratorium on certain types of recombinant DNA experiments and containment on others until the risks were better understood.

The Berg Committee also called for an international meeting of scientists. This led to the 1975 Asilomar Conference organized by the National Academy of Sciences held for four days in Pacific Grove, California. One hundred and forty biologists and physicians, four lawyers, and twelve journalists assembled to discuss the potential risks involved with recombinant DNA technology and to discuss and establish the conditions under which research should proceed.

In the absence of state and federal laws, scientists took it upon themselves to develop a plan on how to proceed with the recombinant DNA technology inside and outside the lab. Scientists suggested NIH form a Recombinant DNA Advisory Committee to establish safety guidelines, standards for conducting experiments, and oversight for NIH funded projects. The scientists preferred the flexibility of guidelines over legislation and regulations since they are less affected by the politics of Washington DC.

For inside the lab, scientists recommended guidelines matching the type of containment necessary for different types of experiments. Experiments categorized as minimal, low, moderate, and high risk would require scientists to follow appropriate safety standards and procedures.

For assessing risks outside the lab, it became apparent that it was necessary to distinguish genetic engineering (a process) from recombinant DNA (the product). One participant of the Asilomar Conference announced that while preparing the guidelines for experimental DNA research, that they had made human sex a moderate risk experiment.[3]

Initially, the city of Cambridge, Massachusetts had banned recombinant DNA research. Following a period without the determination of any real risks, Senator Edward Kennedy said, "Public hysteria cannot be maintained indefinitely in the absence of a credible villain of recombinant DNA technology."[4] We now know genetic engineering is harmless. It is a predictable process and the functions of the transferred genes are known.

Despite many successes with recombinant DNA technology, its applications provide unique situations for policy analysts to consider. Recombinant DNA has a different set of regulatory issues from genetic engineering, including consumer safety, environmental consequences, and potential intentional misuse. At the time, consumer and environmental activists were against scientists self regulating the field in favor of more discussions.

A global coalition of activist organizations led by ETC Group claims civil society representatives were unable to attend the Asilomar Conference because conference officials turned those away who tried to attend due to limited space.[5] Since applications of biotechnologies require public acceptance, if a situation got out of control it could stifle the future development of the field and the potential development of any social goods.

Today, Asilomar provides a model for risk assessment of cutting edge biotechnologies based on three key concepts; oversight through self-regulation by scientists along with government guidelines, determining what is unique about the technology, and addressing public concerns to prevent backlash.

GMOS: A CASE STUDY

Using genetic engineering to recombine DNA has led to many life saving drugs including insulin, EPO, interferon, and human growth hormone. Today, genetic engineering is extremely important in medical research especially for combating deadly influenza viruses.

Scientists have genetically engineered crops to deal with pests. In the 1990s, one of Hawaii's major crops the papaya was in danger of decimation by the ring spot virus. Papayas genetically engineered to resist the virus saved the industry without using herbicides or pesticides.

Each year, increasing numbers of transgenic plants appear in our grocery stores and kitchens. Among the transgenic plants currently approved by the FDA include soybeans, corn, cotton, alfalfa, papaya, rapeseed, and squash. Today, over 80 percent of corn and 90 percent of soybeans grown in America are genetically engineered.

Genetic engineering has also provided a solution to food shortages and nutritional deficiencies in developing countries. Ingo Potrykus developed genetically engineered Golden Rice which includes genes that produce beta-carotene, a precursor to vitamin A, for children in Africa and Southeast Asia.

Futurists are forecasting 9 billion people on earth in 2050; with land, water, and energy restraints. As the world population increases traditional crops may not keep up with food demand, especially in developing nations. Genetically modified organisms (GMOs) can produce higher yields on less acreage at less cost. Even though a number of the third world countries with starving people would benefit from genetically modified crops, some of these countries governments have refused them. In Africa, activists claim the private sector valued profits over the welfare of citizens creating the perception that some biotech companies are greedy and more interested in international trade than environmental and health issues.

The European and Southwestern Corn Borers are responsible for over one billion dollars of damage to American crops annually.[6] When the larvae of corn borers burrow into the corn stalks, traditional chemical insecticides are useless and the crops are lost. The naturally occurring soil bacterium, *Bacillus thuringiensis (Bt)*, produces an insect toxin that eliminates the need for traditional chemical pesticides. Scientists at Monsanto genetically engineered crops to produce their own *Bt* toxins making the crops toxic to moths and butterflies. When the corn borers consume the toxins produced from the *Bt* gene, they are broken into smaller pieces which bind to specific receptors causing an electrolytic imbalance blocking the intake of food. The toxin is not harmful to humans since it is easily degraded in the stomach.

Although farmers have used *Bt* sprays since 1961, *Bt* insecticides have lost effectiveness due to resistance. Insects receiving a sub-lethal dose of *Bt* insecticides may develop resistance. Scientists are investigating several methods to prevent resistance. One strategy is expressing multiple *Bt* toxins or fusing *Bt* toxins together causing multiple mutations, because it is unlikely that insects will evolve resistance to multiple toxins.[7] Since chloroplasts are not carried by pollen, scientists are also investigating introducing genes into chloroplast DNA rather than nuclear DNA.

Exporting genetically engineered crops provides a boost to the biotechnology industry. The negative reaction internationally to genetically engineered crops has impacted American exports. The European Union had a moratorium which banned imports of all genetically modified foods and feed products between June 1998 and August 2003. This cost American farmers roughly \$300,000,000 annually.[8]

As a result of lost revenues, twelve nations filed a suit against the European Union over the moratorium on genetically engineered foods. The World Trade Organization ruled the European Union's GMO moratorium was illegal. In September 2003, participants of an international meeting drafted the Cartagena Protocol which regulates the international transfer of GMOs. This international agreement requires producers of GMOs to provide a detailed risk assessment based on the precautionary principle. Today, applications for licensed GMOs require an extensive risk assessment and regulatory approval process with comments by the public.

Europeans have a less favorable perception of exporting genetically modified organisms (GMOs) than Americans. The mishandling of Mad Cow disease in the 1980s by the British government with a possible cover up may help explain the reaction against genetically engineered crops in Europe.[9]

So, what went wrong with GMOs? In *Rights and Liberties in the Biotech Age*, Krimsky and Shorett argue for the past several decades, society had to adapt to scientific discoveries. Consequently, the burden is placed on concerned citizens to expose risks. They claim science and technology have rights which violate rights of people around the world.

Did public backlash result from the perception that GMO crops are harmful to humans and the environment? Environmentalists were concerned with unintended consequences to humans and ecosystems. GMOs can affect non-target organisms and disruption ecosystems. Understanding potential problems can depend on understanding specific ecosystems and farming practices. It is helpful to know what plants are located near the *Bt* corn and what insects feed on it. *Bt* corn pollen can kill non-target insects as well as the pests for which it is intended. For example, Monarch butterflies do not have the enzyme to process the *Bt* toxin. If farmers use herbicides that kills milkweed, the monarch butterfly's favorite food, the butterflies will move elsewhere to feed or potentially eat *Bt* corn.

Another stakeholder, organic farmers were concerned with losing their organic certification. Since wind blown cross-pollination occurs up to 100 meters away, GMOs may contaminate non-modified crops. It is difficult to totally contain the genetically engineered plant resulting in some contamination from seed stock and cross pollination with the original plant. Despite any attempts by producers to minimize risks, numerous scenarios can occur using GMOs. So, their usage also requires responsible practices by farmers.

In 2010, the FDA began reviewing an application for the first genetically engineered animal food source. AquaBounty has spent approximately $60 million and 15 years engineering a salmon that grows to maturity in 18 months, roughly half the time it takes in the wild. The Atlantic salmon has a modified regulatory gene for growth hormone from the Pacific Chinook Salmon and a promoter from ocean pout.

In a preliminary finding in September 2010, the FDA found the Atlantic salmon safe. The FDA uses a multiple step process which is expensive and time consuming. It is based on the system used for veterinary drugs and to assess genetically engineered animal food sources. The process considers both the potential effects on the environment through the life cycle of the genetically modified fish and its safety as a food source.

In behalf of a coalition of 52 organizations consisting of consumer and environmental groups, the Center for Food Safety is opposing final FDA approval of the Atlantic salmon. Environmental groups are concerned that the genetically engineered Salmon will mix with wild populations. To address the mixing issue, AquaBounty has created females with triploid chromosomes making them sterile, and raises them in isolated tanks.

The Center for Food Safety has questioned the science behind several studies. Consumer groups are concerned with potential allergic reactions from novel expressed proteins. The FDA concluded there is no significant difference between the genetically engineered and wild versions. Also, consumer groups are concerned with the impact of the hormone levels in the bloodstream regarding elevated levels of insulin-like growth factor 1. The FDA has determined that even consuming large quantities of the Atlantic salmon pose no risk.

A federal advisory committee is reviewing for any regulatory gaps and may eventually approve the salmon, but suggest labeling. Consumers can recognize niche products with unique traits; however, it is difficult to distinguish GMOs with increased yield. In the past, some countries have banned GM plant imports if there is not enough scientific evidence to ensure safety. Sellers now provide a detailed description of the product and relevant information that will affect consumer decision making.

If approved, the Atlantic salmon will set a precedent for future approvals. At that point, the FDA will approve genetically engineered animal food sources on a case by case basis. In deciding whether firms selling the wild Atlantic salmon can label their products as non-GE, the FDA will consider if claims are supportable through tracing or testing or if it falsely implies non-GE salmon is safer.[10]

ASILOMAR *REDUX*

A Who's Who of a small but growing group of synthetic biology practitioners regularly meets through a series of conferences to form a cooperative community. While in the earliest stages of synthetic biology development, the 2004 Synthetic Biology 1.0 Conference (SB1) held at MIT brought together researchers. This conference was followed up with SB2 held at Berkeley in 2006, SB3 held at ETH University in Zurich, Switzerland in 2007, SB4 held in Hong Kong in 2008, and SB5 held at Stanford University in 2011.

Among the issues discussed by researchers at these conferences were intellectual property verses sharing of information and the need to adopt a code for self-governance. Similar to the introduction of genetic engineering, bioengineers are self-regulating the field with government guidelines. The current priorities are managing the unknown risks of novel proteins and genomes in the environment and the unique dual use nature of synthetic biology.

BRICOLAGE

Francois Jacob of the Pasteur Institute said nature works as a tinkerer with available materials, not as an engineer does by design.[11] Nature's genetic code is sophisticated, but not necessarily optimized. An example that scientists commonly refer to is the human eye. It is poorly designed i.e. it is wired backwards. Nerves on the retina cause a blind spot. Even though nature has billions of years of evolutionary experience, it had done so with limited resources to work with. With a limitation on the number of building blocks, nature makes use of the enzymes available. Only twenty amino acids are routinely used in building life forms.

If you imagine the three dimensional space for all of the possible configurations of proteins and DNA resulting in life forms, only a very small area of that space is occupied. If evolution depended on nature to produce humans through natural selection, the chances are our species would not exist. Due to evolutionary constraints, millions of years ago symbiosis was necessary for humans to evolve.

If engineers create machines that do not work, they can design a new machine. If the cell gambles with changes that do not work, it fails to generate new cells and all of its information is lost. Since genetic information is passed from generation to generation, cells must maintain a line back to the earliest cells. Engineers have the ability to use technology to improve on nature's designs by overcoming its constraints.

Bioengineers are creating novel proteins and designing protein-protein interfaces. With increased protein stability and altered binding specificity, these artificial proteins with new enzymes to catalyze chemical reactions have potential biomedical applications.

Similar to recombinant DNA technology, it's not biological engineering that is potentially dangerous, rather the organisms that are produced.

Environmental activists are concerned with what effect novel genomes will have on the environment and ecosystems. Will synthetic organisms have enhanced virulence or resistance to vaccines and anti-viral medication?

Will synthetic organisms escape from labs and create havoc? Scientists can create suicide genes so the organism is unable to live outside the lab. For organisms intentionally released into the environment, scientists can create watermarks or genetic signatures which are used to differentiate the synthetic genome from the natural genome and for tracking purposes.

Molecular biologists are attempting to understand what effect the environment will have on novel genomes. Will synthetic organisms mutate or acquire destructive emergent properties? According to Michael Hecht of Princeton University, synthetic enzymes can inadvertently activate cryptic states.[12] However, scientists are uncertain of the activating mechanism.

Based on these concerns, the Bioethics Commission report on synthetic biology released in 2010 urges regulators to focus on a "gap analysis" of existing regulations to identify any missing oversight in current risk assessment practices related to the field release of synthetic organisms.

OPENNESS & BIOSECURITY

The unique dual use nature of synthetic biology has biosecurity implications. There is a risk that someone may use protein engineering to develop designer toxins for military purposes. Of particular concern to security analysts is the generation of hybrid or fusion toxins which are more much more toxic than either single toxin and that are extremely difficult to treat.[13]

Ironically, many of the same activists calling for open source biology are simultaneously expressing concerns regarding the associated biosecurity risks with synthetic pathogens. The shared databases intended for researchers help make the threat of bioterrorism a legitimate concern.

In 2001, Australian scientists reported in *The Journal of Virology* that while developing a contraceptive vaccine to control rodent populations they inserted a gene for an immune system protein into a mousepox virus.[14] This modification unexpectedly made the normally mild virus lethal in mice, even those that were naturally resistant to mousepox or that were vaccinated against it. The scientists received criticism for having the article published because terrorists may want to develop a vaccine resistant strain of poxvirus such as smallpox or monkeypox.

Researchers from the State University of New York at Stony Brook led by Eckard Wimmer aimed to determine the difficulty in synthesizing one of the simplest known viruses, polio. After three years, in 2002, the team synthesized the 7,500 base RNA poliovirus using a published genetic sequence from the internet and customized mail order oligonucleotides, sequences consisting of 50-100 base pairs. After the researchers injected the synthetic virus into mice, they became paralyzed and died.

After the press called the work irresponsible, Wimmer disclosed that the process of synthesizing the 7,741 bases tedious and terrorists would find it much easier to use an existing virus found in nature.[15] Since the polio virus was eradicated in nature, the World Health Organization (WHO) had plans to stop inoculations. The natural polio virus still exists in government labs in the USSR and the United States. Scientists estimate exposure to either the natural or synthesized virus could lead to millions of deaths since humans have little resistance to the virus.

In 2005, scientists determined the sequence of the pathogen responsible for the Spanish influenza virus which killed an estimated 20-50 million people in a pandemic from 1918-1919. From the genetic sequence in a viral sample obtained from a frozen victim found in the Alaskan permafrost, scientists with the United States government were able to synthesize the virus in the laboratory. The scientists tested the synthesized virus on mice to identify the genetic factors which made the strain so deadly and to help researchers develop vaccines and antiviral drugs against future strains of influenza. Critics were not only outraged because of the possibility the virus could escape from the laboratory, but with the CDC's decision to publish the Spanish influenza virus sequence in GenBank.

Unlike many earlier industrial technologies, hobbyists can bioengineer DNA inexpensively and on a small scale. The amateur, do-it-yourself (DIY) community working out of garages is currently able to purchase DNA synthesizers on eBay for approximately $9,000. Now that scientists have established a proof of principle for creating a deadly virus from genetic code which is available in public databases, the DIY community proposes a potential biosecurity risk.

In a study weighing the advantages and disadvantages of restricting access, the National Research Council concluded that there is no guarantee terrorists could recreate a lethal pathogen.[16] This study was released in 2004 before the recent development of DNA printing technology. Low cost, high throughput on chip gene synthesis could change that scenario.[17] Appropriately, as the field develops further, the Bioethics Commission's motto moving forward is "prudent vigilance."[18]

In a report financed by the Alfred P. Sloan Foundation, The Venter Institute teamed up with MIT and The Center for Strategic and International Studies to help develop guidelines in order for synthetic biology to proceed safely. This report concluded that in most cases it is easier to steal from a laboratory or order DNA sequences online than to synthesize them.[19] Given the complexity of biological systems, it remains difficult to create synthetic pathogens.

DNA synthesis companies provide oligonucleotides primarily to researchers for basic research and to develop vaccines against diseases. In 2006, James Randerson, a science reporter with *The Guardian*, investigated how easy it is to order a pathogen's genetic code which is available from an online database. For safety reasons, he sought advice from a synthetic biologist to make sure the partial sequence was not dangerous.[20] He chose a 78 letter DNA sequence that is an analog to a sequence in the smallpox virus and did not code for a virulent gene. It also had three nucleotide modifications.[21]

In 2006, he ordered a partial sequence of the smallpox virus from a British sequencing company and had it delivered to a residential address. With all the orders they receive daily it is difficult to screen all of them. The supplier was not aware the sequence coded for a destructive organism.

As part of self-governance, the synbio community has attempted to close this loophole using several methods. Scientists have developed a Code of Conduct for Best Practices of Gene Synthesis to screen sequences. Companies that sell DNA sequences can cooperate through a screening process. Blackwatch, a software developed at Craic Computing, allows companies to screen for oligonucleotides and compare the DNA sequences to a known list of pathogens.

Bioengineers have organized an International Gene Sequencing Consortium to restrict evildoers from purchasing pathogen sequences. Companies can allow only those with professional certification and secure permits to place orders. Some bioengineers are boycotting DNA synthesizing companies that do not screen their orders and work only with companies that fill orders to qualified people.

A DIFFERENT STANDARD?

Similar to genetic engineering which has produced so many social goods, ETC Group is against scientist's self-regulating synthetic biology. Jim Thomas of ETC Group told the Bioethics Commission that the position of the coalition of civil society activists is to ban synthetic biology research.[22] The coalition of civil society groups has called for global governance rather than relying on industry self-regulation. As a second option, ETC Group is asking for a moratorium on research until more open debate takes place, the pre-cautionary approach, which gives the public a chance to thoroughly discuss all the potential benefits as well as potential risks to stakeholders before biotechnologies are used.[23]

ETC Group has recommended a ban on the release of synthetic organisms into the environment until there is global governance and more discussions occur. In a news release, ETC Group argues the oversight of environmental risks is inadequate by placing faith in suicide genes.[24] Also, the activists claim the partnering of BP and Exxon with Craig Venter's Synthetic Genomics for commercialization of synthetic fuels is irresponsible and potentially environmentally damaging.[25]

However, as a society, Americans are not Luddites, we want progress and to improve the quality of our lives. Accordingly, in his letter to the Bioethics Commission, President Obama expressly states that the task is to develop recommendations that will allow citizens to reap the benefits of this industrial revolution such as accelerated vaccine development.[26] There is no mention in the letter of deep philosophical discussions which could potentially sidetrack their discussions.

Following the 2001 anthrax attacks in the United States, officials arranged for the manufacture of a vaccine to protect against polio and other lethal pathogens, and the World Health Organization is also planning to stockpile vaccines. If bioterrorists were to release lethal pathogens into the air, water, or food supply, security analysts stress the need for the development of the next generation of countermeasures.[27] This, ironically, is best accomplished by developing the field of synthetic biology. During testimony to the Bioethics Commission, reflecting Obama's sentiments bioengineer Drew Endy said we need to figure out how to get this right.[28]

To ensure citizens safely receive the social benefits from synthetic biology, ideally oversight would take place through a rigorous process similar to clinical trials that the pharmaceutical industry uses for drugs. However, clinical trials are very expensive and subsequently cost prohibitive to the development of most industrial products. For example, providing environmental impact statements for every possible scenario using synthetic proteins and genomes is impossible. Besides, in spite of the costs, a study reported in the Journal of the American Medical Association reveals that roughly 106,000 people die each year in American hospitals as the result of side effects from medication.[29]

For the regulation of automobile safety, Americans have a social contract based on cost-benefit analysis. According to the National Highway Traffic Safety Administration, over the last twenty years Americans have accepted roughly 40,000 traffic fatalities annually in return for a convenience that is engrained as part of our lifestyle.[30] If the speed limit were lowered to 45 mph, it would undoubtedly result in fewer fatalities. But, how many activists or members of Congress would agree to lower the speed limit on highways to 45 mph?

While automobiles and pharmaceuticals are heavily regulated, human fatalities are common and tolerated. For self-governed biotechnologies including genetic engineering, nanotechnology, and synthetic biology fatalities are non-existent. Some technologies with known risks have societal acceptance while others do not. Are activists are holding synthetic biology to a different standard?

Notes and Sources

EPIGRAPH

1. Diana Schaub. How to Think about Bioethics and the Constitution. American Enterprise Institute. 1. 2004.
2. Jean Baptiste Pierre Lamarck 1809. *Philosophie Zoologique*. Dentu.
3. August Weismann. 1893. The Germ-Plasm: A Theory of Heredity. Charles Scribner's Sons. 183.
4. Sahotra Sarkar. 1996. The Philosophy and History of Molecular Biology: New Perspectives. Kluwer Academic Publishers. 202.
5. Francois Jacob. Evolution and Tinkering. *Science*. 196(4295): 1161-1166. 1977

CHAPTER 1 WHAT GENOMICS REVOLUTION?

1. Randall Mayes. A Review of J. Craig Venter's A Life Decoded. *Journal of Evolution and Technology*. 71-72. 2008. http://jetpress.org/v17/mayes.htm
2. Colette Dib et al. A comprehensive genetic map of the human genome based on 5,264 microsatellites. *Nature*. 380:152-154.
3. R. Dulbecco A Turning Point in Cancer Research: Sequencing the Human Genome. *Science*. 231:1055-56. 1986.
4. Phone interview with James Wyngaarden, October 4, 2005
5. John R. Inglis, Joseph Sambrook, and Jan A. Witkowski. 2003. Inspiring Science: Jim Watson and the Age of DNA. Cold Spring Harbor Laboratory Press. 377-379.
6. Phone interview with James Wyngaarden, October 6, 2005
7. Inglis, Sambrook, and Witkowski. 2003. 377-379.
8. Venter. 2007. 137.
9. J. Craig Venter. 2007. A Life Decoded. Viking Penguin. 243.
10. James Shreeve. 2004. The Genome War. Alfred A. Knopf. 8-11.
11. Victor McElheny. 2010. Drawing the Map of Life: Inside the Human Genome Project. Basic Books. 146, 154.
12. Philip Benfey and Alexander Protopapas. 2004. Essentials of Genomics. Prentice Hall. 108.

13. Venter. 2007. 321.

14. Llewellyn H. Rockwell. Jr. Government's Genetic Failure, Ludwig von Mises Institute. June 27, 2000. http://www.mises.org/story/455

15. Robert Carlson. The Pace and Proliferation of Biological Technologies. *Biosecurity and Bioterrorism: Biodefense Strategy, Practice, and Science.* 2003. 1(3):4.

16. Venter. 2007. 247-249

17. John Sulston and Georgina Ferry. 2002. The Common Thread. Joseph Henry Press. 160.

18. Geoffrey Carr. Biology 2.0. *The Economist.* June 17, 2010 http:// economist.com/node/16349358

19. Peer Bork and Martijn Huynen. A Point of Entry into Genomics. *Nature Genetics.* 23:273. 1999.

20. S. Anderson et al. Sequence and Organization of the Human Mitochondrial Genome. *Nature.* 290:457-465. 1981.

21. George L. Kenyon. 2002. Defining the Mandate of Proteomics in the Post-Genomics Era: Workshop Report. National Academy of Sciences. 766.

22. Mary Chitty. Cambridge Healthcare Institute. 2008. http://www.genomicglossaries.com/content/genomic_overview.asp

23. http://www.dictionary.com

24. J. R. Riordan et al. Identification of the cystic fibrosis gene: cloning and characterization of complementary DNA. *Science.* 245(4922):1066-1073. 1989.

25. Benfey and Protopapas, 2005. 518-519.

26. Ibid.

27. http://www.ncbi.nlm.nih.gov/pubmed; http://www.ncbi.nlm.nih.gov/genbank

CHAPTER 2 MINING GENOMES

1. Jan Klein and Akie Sato. The HLA System: Part One. *The New England Journal of Medicine.* 343(10):702. 2000.

2. Christopher Wills. 1991. Exons, Introns, and Talking Genes. Harper-Collins. 278.

3. Personal Genome Project. http://www.personalgenomes.org/mission.html

4. 1000 Genomes. A Deep Catalog of Human Genetic Variation. http://1000genomes.org

5. Philip Benfey and Alexander Protopapas. 2005. Genomics. Prentice Hall. 560.

6. Diana Schaub. How to Think about Bioethics and the Constitution. American Enterprise Institute. 1. 2004.

7. Sheldon Krimsky and Peter Shorett. 2005. Rights and Liberties in the Biotech Age. Rowan & Littlefield. 223-224.

8. Claude Moore Health Science Library. Carrie Buck, Virginia's Test Case. 2008. www.http//www.hsl.virginia.edu/historical/eugenics/3-buckvbell.cfm

9. David P. Mindell. 2006. The Evolving World. Harvard University Press. 274.

10. US apologizes for infecting Guatemalans with STDs in the late 1940s. October 1, 2010. http://www.cnn.com

11. U.S. Supreme Court. TVA v. HILL (437 U.S. 153). Argued April 18, 1978. 8. http://caselaw.lp.findlaw.com/scripts/getcase.pl?court=us&vol=437&invol=153

12. James Drozdowski. Saving an endangered act: the case for a biodiversity approach to ESA conservation efforts. Case Western Reserve Law Review. 45(2):569. 1995.

13. Department of Health and Human Services. Code of Federal Regulations. Part 46. Protection of Human Subjects http://hhs.gov/ohrp/humansubjects/guidance/45cfr46.htm

14. Philip Bereano. Patent Pending: The Race to Own DNA. *Seattle Times.* B5. August 27 1995.

15. Marina L. Whelan. What if any, are the Ethical Obligations of the U.S. Patent Office? *Duke Law & Technology Review*. 0014:6. 2006.

16. Spencer Wells. 2006. Deep Ancestry. National Geographic Society. 170-174.

17. Leslie E. Wolf. Advancing Research on Stored Biological Materials: Reconciling Law, Ethics, and Practice. *Minnesota Journal of Law, Science, and Technology*. 11(1):115-116. 2010.

18. Christopher Jackson. Learning from the Mistakes of the Past: Disclosure of Financial Conflicts of Interest and Genetic Research. *Richmond Journal of Law and Technology*. 2004. Volume XI. Issue 1. http://law.richmond.edu/jolt/v11i1/article4.pdf

19. Wolf. 2010. 105-107

20. Amy Harmon. Indian Tribe Wins Fight to Limit Research on Its DNA. *The New York Times*. April 21, 2010.

21. Paul Rubin. Indian Givers. *Phoenix New Times*. May 27, 2004.

22. David E. Winickoff. Governing Populations Genomics: Law, Bioethics, and Biopolitics in Three Case Studies. *Jurimetrics*. 43:201-203. 2003.

23. Neil A. Manson and Gregory Conko. Genetic Testing and Insurance: Why the Fear of Genetic Discrimination does not Justify Regulation. Competitive Enterprise Institute. Issue Analysis No.4. April 5, 2007.

24. Thomas Boyden. New Law Trashes Genetic Science. The Ayn Rand Institute Press Release. June 11, 2008.

25. Llewellyn H. Rockwell, Jr., Government's Genetic Failure. Ludwig von Mises Institute. 2000. http://www.mises.org/story/455

26. Neal Dickert and Jeremy Sugarman. Getting the Ethics Right Regarding Research in the Emergency Setting: Lessons from the PolyHeme Study. *Kennedy Institute of Ethics Journal*. 17(2):162-163. 2007.

27. Karen Kaplan. U.S. military practices genetic discrimination in denying benefits. *Los Angeles Times*. August 18, 2007.

28. Alex Hutchinson. Astronaut Love Triangle Highlights Mars Mission Challenge: Avoiding Crazy on Long Flights. *Popular Mechanics*. October 2009. http://www.popularmechanics.com/science/4212593

29. F. A. Cucinotta et al. Space Radiation and Cataracts in Astronauts. *Radiation Research*. 156:460-466. 2001.

CHAPTER 3 OUT OF THE WOODWORK

1. Randall Mayes. In Defense of Patenting DNA: A Pragmatic Libertarian Perspective. Institute for Ethics & Emerging Technologies. 2009. http://ieet.org/index.php/IEET/more/mayes20090726

2. Ibid.

3. Maurice Cassier. Appropriation and Commercialization of the Pasteur Anthrax Vaccine. *Studies in History and Philosophy of Science*. 36(4):1. 2005.

4. Discovery of Penicillin. American Chemical Society. http://acswebcontent.acs.org/landmarks/penicillin/index.html

5. Lila Feisee. Anything Under the Sun Made by Man. 2001. Biotechnology Industry Association. http://bio.org/speeches/speeches/041101.asp

6. Donald Stokes. 1997. Paster's Quadrant: Basic Science and Technological Innovation. Brookings Institution Press.

7. Steven Burrill. 2003. Biotech Life Sciences: Revaluation and Restructuring: 17th Annual Report on the Industry. Burrill & Company. 338.

8. J. A. DiMasi et al. The Price of Innovation: New Estimates of Drug Development Costs. *Journal of Health Economics.* 22:180. 2003.
9. Michael S. Mireles. An Examination of Patents, Licensing, Research Tools, and the Tragedy of the Anticommons in biotechnology Innovation. *University of Michigan Journal of Law Reform.* 38(162):144. 2004.
10. Catherine Peck. A Century at Stanford. *Stanford Magazine.* Nov./Dec. 1998.
11. Andrew Pollack. Amgen Wins Anemia Drug Patent Battle with Roche. *The New York Times.* C2. October 24, 2007.
12. John Sulston and Georgina Ferry. 2002. The Common Thread. Joseph Henry Press. 202, 266.
13. Jack Wilson. No Patents for Semantic Information. *The American Journal of Bioethics.* 2(3):15-16. 2002.
14. J. Craig Venter. 2007. A Life Decoded. Viking Penguin. 130.
15. David Koepsell. 2009. Who Owns You?: The Corporate Gold-Rush to Patent Your Genes. Wiley-Blackwell. 5.
16. John Schwartz. Cancer Patients Challenge the Patenting of a Gene. *The New York Times.* May 13, 2009.
17. Joy Yang. Our Bodies,Whose Patents? *The Daily Beast.* 2010. http://thedailybeat.com/newsweek/2010/11/14/who-owns-our-genes.htm
18. New Terminator Patent goes to Sygentia. ETC Group. 2001. http://www.etcgroup.org/en/node/277
19. ICTA Analysis of Supreme Court Decision in Patent Case. International Center for Technology Assessment. 2001. http://www.cropchoice.com/leadstrya594.html?recid=540
20. Andrea Gawrylewski. USPTO upholds stem cell patent. 2008. http://the-scientist.com/blog/display/54389
21. E-mail from Daniel Kevles. August 11, 2005.
22. Sharyl J. Nass and Bruce W. Stillman. 2003. Large-Scale Biomedical Science. National Academy of Sciences. 165.
23. M. A. Heller and R. Eisenberg. Can Patents Deter Innovation? The Anticommons in Biomedical Research. *Science.* 280(5364):698. 1998.
24. Ted Buckley. The Myth of the Anticommons. Biotechnology Industry Association. 13. 2007. http://www.bio.org/ip/domestic/TheMythoftheAnticommons.pdf
25. Buckley. 2007. 2.
26. John P. Walsh, Ashish Arora, and Wesley M. Cohen. 2003. Effects of Research Tool Patents and Licensing on Biomedical Innovation. National Academy of Sciences. http://books.nap.edu/openbook.php?isbn=0309086361&page=285
27. National Academy of Sciences. 2005. Reaping the Benefits of Genomic and Proteomic Research: Intellectual Property Rights, Innovation, and Public Health. The National Academies Press. http://www.nap.edu/catalog/11487.html
28. Ted Hagelin. The Experimental Use Exemptions to Patent Infringement. New York State Science & Technology Law Center. 11. 2005.
29. Hagelin. 2005. 6.
30. Hagelin. 2005. 14-15.
31. Hagelin, 2005. 18-19.
32. Randall Mayes. Book Review of David Koepsell's Who Owns You?: The Corporate Gold Rush to Patent your Genes. Journal of Evolution and Technology. 2009. http://jetpress.org/v20/mayes.htm

33. Scott Kieff. Who Owns Your Genes? National Press Club transcript. July10, 2007. http://www.DNApolicy.org/resources/PatentingGenePOPStranscript.pdf

CHAPTER 4 THE RISE & DESCENT OF VITALISM

1. Jean Baptiste Pierre Lamarck. 1809. *Philosophie Zoologique*. Dentu.
2. August Weismann. 1893. The Germ-Plasm: A Theory of Heredity. Charles Scribner's Sons. 183.
3. Peter J. Bowler. 2003. Evolution: The History of an Idea. University of California Press.
4. Maitland A. Edey and Donald C. Johanson. 1989. Blueprints: Solving the Mystery of Evolution. Little, Brown and Company. 70-72.
5. Robert Chambers. 1969. Vestiges of the Natural History of Creation. Humanities Press. 33. (Leicester University Press published the first edition in 1844)

CHAPTER 5 HOW THE GIRAFFE GOT ITS NECK

1. Thomas Kuhn. 1962. The Structure of Scientific Revolutions. University of Chicago Press. 11-12.
2. Ernst Mayr. 2004. What Makes Biology Unique? Cambridge University Press. 159-169.
3. Ernst Mayr. 80 Years of Watching the Evolutionary Scenery. *Science*. 305(5680):46-47. 2004.
4. Peter J. Bowler 1988. The Non-Darwinian Revolution: Reinterpreting a Historical Myth. Johns Hopkins U. Press. 3.
5. Mayr. 2004. 159-169.
6. Peter J. Bowler. 1983. The Eclipse of Darwinism. Johns Hopkins University Press.
7. Robert Olby. Horticulture: The Font for the Baptism of Genetics. *Nature Reviews Genetics*. 1:65-70. October 2000.
8. James D. Watson 2004. *DNA: The Secret of Life*. Alfred A. Knopf. 310.
9. Brian K. Hall. Unlocking the Black Box between Genotype and Phenotype. *Biology and Philosophy*. 18(2):222-225. 2003.
10. Robert J. Richards 1992. The Meaning of Evolution. University of Chicago Press. 57.
11. Hall. 2003. 222-225.
12. Greg Wray. Resolving the *Hox* Paradox. *Science*. 292(5525):2256. 2001.
13. Sahotra Sarkar. 1996. The Philosophy and History of Molecular Biology: New Perspectives. Kluwer Academic Publishers. 202.
14. Hall. 2003. 222-225.
15. Rudolph A. Raff. Evo-devo: The Evolution of a New Discipline. *Nature Reviews Genetics*. 1:75. 2000.
16. Sean B. Carroll, Nicholas Gompel, and Benjamin Prudhomme. Regulating Evolution: How Gene Switches Make Life. *Scientific American*. May 2008.
17. Wray. 2001. 2256.
18. University of Chicago Medical Center Press Release. Evolution Re-Sculpted Animal Limbs by Genetic Switches Once Thought too Drastic for Survival. *ScienceDaily*. August 18, 1997.
19. Carroll, Gompel, and Prudhomme. 2008.
20. Sean Carroll. 2005. Endless Forms Most Beautiful. W.W. Norton. 118-122.
21. E-mail correspondence from Greg Wray of Duke University on September 8, 2011.
22. Michael Ruse. The Darwinian Revolution as seen in 1979 and as seen Twenty-Five Years Later in 2004. *Journal of the History of Biology*. 38(1):6-13. 2005.

23. John F. McCarthy. Is The Genesis Account of Creation Literally True? 2005. http://www.rtforum.org/lt/lt120.html
24. District Court Judge John Jomes' Decision for *Kitzmiller v. Dover Area School District*. 32. December 20, 2005. http://www.pamd.uscourts.gov/kitzmiller/kitzmiller_342.pdf
25. Michael. Behe. 1996. Darwin's Black Box. The Free Press. 26.
26. Michael J. Behe. 2007. The Edge of Evolution. The Free Press. 186.
27. The Discovery Institute. The Wedge Document, So What? 1999. http://discovery.org/a/2101
28. Laura Beil. Opponents of Evolution Adopting a New Strategy. *The New York Times*. June 4, 2008.
29. Jeff. Hecht. Why Doesn't America Believe in Evolution? *New Scientist*. 191(2565):11. 2006.
30. Nothing in Biology Makes Sense Except in the Light of Evolution. WGBH Educational Foundation. 2001. http://www.pbs.org/wgbh/evolution/library/10/2/text_pop/l_102_01.html

CHAPTER 6 FOUR WAVES

1. Evelyn Fox Keller. 2000. The Century of the Gene. Harvard University Press. 8.
2. Timothy Lenoir. Revolution from Above: The Role of the State in Creating the German Research System, 1810-1910. *The American Economic Review*. 88(2):22. 1988.
3. Walter S. Sutton. The Chromosomes in Heredity. *Biological Bulletin*. 4:231-251. 1903.
4. Oswald T. Avery, Colin M. MacLeod, and Maclyn McCarty. Studies on the Chemical Nature of the Substance Inducing Transformation of Pneumococcal Types. *The Journal of Experimental Medicine*. 79(2):137-158. 1944. http://jem.rupress.org/content/79/2/137.full.pdf.html
5. Lawrence Altman. Maclyn McCarty Dies at 93; Pioneer in DNA Research. *The New York Times*. January 6, 2005.
6. Martin Hewlett. From Mendel to Biotechnology. 1997. http://www.mcb.arizona.edu/Hewlett/mjhpaper.html
7. Hewlett. 1997.
8. Scott Freeman. 2002. Biological Science. Prentice Hall. 61.
9. Maitland A. Edey and Donald C. Johanson. 1989. Blueprints: Solving the Mystery of Evolution. Little, Brown and Company. 226.
10. J. C. Venter. Testimony before the Subcommittee on Energy and Environment. U.S. House of Representatives Committee on Science. June 17, 1998.
11. Francis S. Collins et. al. US National Genome Research Institute, A Vision for the Future of Genomics Research. *Nature*. 422:6934. 2003.
12. BBC News. Nature or Nurture? February 11, 2001. http://news.bbc.co.uk/2/hi/science/nature/1164792.stm
13. International Human Genome Sequencing Consortium. Finishing the Euchromatic Sequence of the Human Genome. *Nature*. 431(21):943. 2004.
14. Craig J. Venter et al. The Sequence of the Human Genome. *Science*. 291(5507):1323. 2001.
15. Jennifer F. Hughes and John M. Coffin.. Evidence for Genomic Rearrangements Mediated by Human Endogenous Retroviruses during Primate Evolution. *Nature Genetics*. 29(4):487. 2001.
16. F. Jacob and J. Monad. Genetic Regulatory Mechanisms in the Synthesis of Proteins. *Journal of Molecular Biology*. 3:318-356. 1961.
17. Anthony Griffiths. 1999. Modern Genetic Analysis. W.H. Freeman. 382.

18. Judy Lieberman. Master of the Cell. *The Scientist*. 2010. http://www.the-scientist. com/article/display/57249

19. Gregory J. Hannon. 2003. RNAi: A Guide to Gene Silencing. Cold Spring Harbor Laboratory Press. 75.

20. David A. Lawrence and Eric J. Alm. Rapid Evolutionary Innovation during an Archean Expansion. *Nature*. 469(7328):93-96. 2011.

21. National Institutes of Health. New Genome Comparison Finds Chimps, Humans Very Similar at the DNA Level. 2005.

22. Nicholas Wade. Signs of Neanderthals Mating with Humans. *The New York Times*. May 6, 2010.

23. Jonathan Marks. 2002. What it Means to be 98% Chimpanzee. University of California Press. 81.

24. Rebecca L. Cann, Mark Stone King, and Allan C. Wilson. Mitochondrial DNA and Human Evolution. *Nature* 325: 31-36. 1987.

25. Spencer Wells. 2002. The Journey of Man. Random House. 180-181.

CHAPTER 7 CROSSING THE WEISMANN BARRIER

1. Evelyn Fox Keller and Elisabeth A. Lloyd. 1992. Keywords in Evolutionary Biology. Harvard University Press. 191.

2. Mae-Wan Ho. 1998. Genetic Engineering: Dream or Nightmare. Gateway Books. 129.

3. Interview with Randy Jirtle of Duke University, September 14, 2004

4. R. Holiday and J.E. Pugh. DNA Modification Mechanisms and Gene Activity during Development. *Science*. 187(4173):226-232. 1975.

5. Leslie A. Pray. Epigenetics: Genome, Meet Your Environment. *The Scientist*. 18:13. July 2004.

6. Alexandra Chittka and Lars Chittka. Epigenetics of Royalty. PloS Biology 8(11). e10005320. 2010.

7. Robert A. Waterland and Randy L. Jirtle. Transposable Elements: Targets for Early Nutritional Effects on Epigenetic Gene Regulation. *Molecular and Cell Biology*. 23(15):5293-5300. 2003.

8. M. F. Lyon. Gene Action in the X chromosome of the Mouse. *Nature* 190(4773):372. 1961.

9. Ricki Lewis. Solid Gold Sheepstakes. *The Scientist*. October 2002. http://www. f1000scientist.com/articles/display13337

10. J. Hawks, E. Wang, G. Cochran, H. Harpending, and R. Moyzis. Recent Acceleration of Human Adaptive Evolution. PNAS. 104(52):20753. 2007.

11. John F. Odling-Smee, Kevin N. Laland, and Marcus W. Feldman. 2003. Niche Construction: The Neglected Process in Evolution. Princeton University Press. 21, 30.

12. Spencer Wells. 2006. Deep Ancestry. National Geographic Society.

13. P. C. Sabeti et al. Positive Natural Selection in the Human Lineage. *Science*. 312:1614. 2006.

14. Spencer Wells. 2002. The Journey of Man. Random House. 158-159.

15. K. N. Laland, J. Odling-Smee, and M. W. Feldman. Cultural Niche Construction and Human Evolution. *Journal of Evolutionary Biology*. 14:22-33. 2001.

16. E. Wang, G. Kodama, P. Baldi, and R. Moyzis. 2006. Global Landscape of Recent Inferred Darwinian Selection for *Homo Sapiens*. National Academy of Sciences. 103(1):135. http://www.pnas.org/cgi/doi/10.1073/pnas.0509691102

17. Brian Mattmiller. Genome Study Places Humans in Evolutionary Fast Lane. 2007. www.news.wisc.edu/14548

CHAPTER 8 THE GENOMICS BUBBLE

1. R.M. Cook-Deegan. Origins of the Human Genome Project. *Risk: Health, Safety & Environment.* Vol.5. 97–118. Spring 1994. http://www.piercelaw.edu/risk/vol5/spring/cookdeeg.htm
2. Robert Carlson. 2010. Biology is Technology. Harvard University Press. 150-151.
3. Carlson. 2010. 150-151.
4. Carlson. 2010. 150-151.
5. Battelle Technology Partnership Practice. 2001. Economic Impact of the Human Genome Project. 5-15. http://www.battelle.org/publications/humangenomeproject.pdf
6. Mayo Clinic Staff. Pharmacogenomics: When drug treatment becomes personalized medicine. 2008. http://www.mayoclinic.com/health/personalized-medicine/CA00078
7. Sahotra Sarkar. Post-Genomics. MedBioWorld. 2005. http://medbioworld.com/postgenomics_blog?p=12
8. Alan Aderem. Systems Biology: Its Practice and Challenges. *Cell.* 121(4):511-513. 2005.
9. Tom Reynolds. Genome Data Announcement Fuels Stock Plunge, Misunderstanding. *Journal of the National Cancer Institute.* 92(8):594-597. 2000. http://jnci.oxfordjournals.org/content/92/8/594
10. Misha Angrist and Robert M. Cook-Deegan. Who Owns the Genome? *The New Atlantis.* 87-96. Winter 2006.
11. Marcus Lillkvist. Investigating the Performance of Research Companies. *Innovation Journalism.* Vol. 1 No. 4. 2004.
12. Dan Vorhaus. What Five FDA Letters Mean to the Future of DTC Testing. *Genomics Law Report.* June 2010.
13. Amy Maxmen. Web Genomics Exposes Ethics Gap. *The Scientist.* 2010. http://the-scientist.com/news/display/57512
14. U.S. Food and Drug Administration. 2004. Innovation or Stagnation: Challenge and Opportunity on the Critical Path to New Medical Products. 10.
15. Robert Goldberg and Peter Pitts. Prescription for Progress: The Critical Path to Drug Development. FDA Task Force. 1-10. June 2006. http://www.manhattan-institute.org/html/fda_task_1.htm
16. Organization for Economic Co-Operation and Development. 2009. The Bioeconomy to 2030: Designing a Policy Agenda. 69.
17. President's Council of Advisors on Science and Technology. 2008. Priorities for Personalized Medicine. 30.
18. National Human Genome Research Institute. Electronic Medical Records and Genomics Network. http://www.genome.gov/27540473
19. Francis S. Collins. The Case for a US Prospective Cohort Study of Genes and Environment. *Nature.* 429:475-476. 2004.
20. J. Lazarou, BH Pomeranz, and PN Corey. Incidence of adverse drug reactions in hospitalized patients: a meta-analysis of prospective studies. JAMA. 279(15):1200-1205. 1998.
21. USFDA. 2004. 1.
22. Gardiner Harris. Federal Research Center Will Help Develop Medicines. *The New York Times.* January 22, 2011.
23. R. L. Woolsey and J. Cossman. Drug Development and the FDA's Critical Path Initiative. *Clinical Pharmacology and Therapeutics.* 81(1):129. 2007.
24. Erik Vance. New Medicine Means Research Rethink. *The Scientist.* 2009. http://the-scientist.com/news/display/55748

25. Marion E. Glick. Leptin Helps Body Regulate Fat, Links to Diet. 1995. http://www. rockefeller.edu/pubinfo/leptinlevel.nr.html
26. Glick. 1995.
27. Vijay Yadav et al. A Serotonin-Dependent Mechanism Explains the Leptin Regulation of Bone Mass, Appetite, and Energy Expenditure. *Cell.* September 2009.
28. Avshalom Caspi et al. Influence of Life Stress on Depression: Moderation by a Polymorphism in the *5-HTT* Gene. *Science.* 18(301):386. 2003
29. Andrew Pollack. Drugmaker's Fever for the Power of RNA Interference has Cooled. *The New York Times.* D1. February 7, 2011.
30. Tia Ghose. The RNA Roots of Obesity? *The Scientist.* 2011. http://the-scientist. com/2011/06/08/the-rna-roots-of-obesity
31. Yi Li. Epigenetics: Heritable Changes in Gene Expression.*Vertices.* Duke University Journal of Science and Technology. Vol. XX, No.1. 18. Spring 2004.
32. Ronald Bailey. Tallying the New Bioethics Council: Has Leon Kass Stacked the Deck? *Reason.* 2002. http://reason.com/rb/rb012302.shtml
33. Boyce Rensberger. Microbes are Immortal, So Why aren't Humans? *The Washington Post.* H1. June 10, 1998.
34. Bruno de Jesus et al. The telomerase activator TA-65 elongates short telomeres and increases health span of adult/old mice without increasing cancer incidence. *Aging Cell.* 1-18. April 2011.
35. Nicholas Wade. Cell Rejuvenation may Yield Rush of Medical Advances. *The New York Times.* January 20, 1998.

CHAPTER 9 WHAT CAN ECONOMISTS LEARN FROM EVOLUTIONARY BIOLOGY?

1. Thomas Woods Jr. 2009. Meltdown. Regnery Publishing. 13-32.
2. Gretchen Morgenson and Joshua Rosner. 2011. Reckless Endangerment. Times Books. 183.
3. Simon Johnson and James Kwak. 2010. 13 Bankers. Pantheon Books. 82.
4. Interview with Michael Greenberger. 2009.The Warning. *Frontline.* PBS. http://pbs. org/wgbb/pages/frontline/warning/interviews/greenberger.html
5. Morgenson and Rosner. 2011. 205.
6. The Warning. 2009. *Frontline.* PBS. 11. http://pbs.org/wgbb/pages/frontline/ warning/etc/script.html
7. Interview with Brooksley Born. 2009.The Warning. *Frontline.* PBS. http://pbs.org/ wgbb/pages/frontline/warning/interviews/born.html
8. Ibid.
9. Roy E. Cordato. Market-Based Environmentalism and Free Market: They're not the Same. *The Independent Review.* 1(3):371-374. Winter 1997.
10. Interview with Brooksley Born. 2009.The Warning. *Frontline.* PBS. http://pbs.org/ wgbb/pages/frontline/warning/interviews/born.html
11. The Warning. 2009. *Frontline.* PBS. 14. http://pbs.org/wgbb/pages/frontline/ warning/etc/script.html
12. Guy Sorman. 2009. Economics Does Not Lie: A Defense of the Free Market in a Time of Crisis. Encounter Books. 1
13. Peter Schweizer. 2009. Architects of Ruin. Harper-Collins. 167-184.
14. Sorman. 2009. 209-210.

15. Paul Krugman. What Economists Can Learn from Evolutionary Theorists. 2. 1996. http://web.mit.edu/krugman/www/evolute.html

16. Paul Krugman. The Power of Biobabble. *Slate.* 1997. http://www.slate.com/id/1925

17. S. J. Gould and R. C. Lewontin. The Spandrels of San Marco and the Panglossian Paradigm: A Critique of the Adaptationist Programme, Proceedings of the Royal Society of London. B205:583-593. 1979.

18. Massimo Pigliucci and Jonathan Kaplan. The Fall and Rise of Dr. Pangloss: Adaptationism and the Spandrels Paper 20 Years Later. *Trends in Ecology and Evolution.* 15(2):66. 2000.

19. Gould and Lewontin. 1979. 2.

20. Stephen Jay Gould. 2002. The Structure of Evolutionary Theory. Harvard University Press. 1252-1253.

21. Niles Eldredge and Steven Gould. 1972. Punctuated Equilibrium: An Alternative to Phyletic Gradualism. (in Models of Paleobiology. T Schopf editor). Freeman, Cooper and Co. 82-115.

22. Lynn Margulis and Dorian Sagan. 1997. Slanted Truths. Copernicus Publishing. 40.

23. Siv G. E. Andersson et al. The genome sequence of *Rickettsia prowazekii* and the origin of mitochondria. *Nature.* 396:133-140. 1998.

24. Randall Mayes. Book review: Robert Carlson's *Biology is Technology: The Promise, Peril, and New Business of Engineering Life. Journal of Evolution and Technology.* Vol. 21. Issue 1. 55-59. 2010. http://jetpress.org/v21/mayes.htm

25. Joseph A. Schumpeter. 2009. Can Capitalism Survive?: Creative Destruction and the Future of the Global Economy. Harper-Collins. 38-47.

26. Sorman. 2009. 94.

27. Executive Office of the President. 2009. A Strategy for American Innovation: Driving Towards Sustainable Growth and Quality Jobs. 6.

28. Executive Office of the President. 2011. A Strategy for American Innovation: Securing Our Economic Growth and Prosperity. Introduction. 2-3.

29. Sorman. 2009. 93-107.

30. Executive Office of the President. 2009. 17.

31. Executive Office of the President. 2009. 15-16.

CHAPTER 10 THE FREE-RIDER PROBLEM

1. Victoria Hale, Jay Keasling, Neil Renninger, and Thierry Diagana. Microbially Derived Artemisinin: A Biotechnology Solution to the Global Problem of Access to Affordable Antimalarial Drugs. *American Journal of Tropical Medicine and Hygiene.* 77:198-202. 2007.

2. Hale, Keasling, Renninger, and Diagana. 2007. 198-202.

3. Clifford Winston. 2006. Government Failure versus Market Failure. AEI-Brookings Joint Center for Regulatory Studies. 61.

4. H. G. Khorana et al. Total Synthesis of the Structural Gene for an Alanine Transfer Ribonucleic Acid from Yeast. *Journal of Molecular Biology* 72:209-217. 1972

5. Elizabeth Corcoran. Stalking a Killer. *California Magazine.* Vol. 117. No. 6. Nov./ Dec. 2006. http://alumni.berkeley.edu/calmag/200611/corcoran.asp

6. F. Jacob and J. Monad. Genetic Regulatory Mechanisms in the Synthesis of Proteins. *Journal of Molecular Biology.* 3:318-356. 1961.

7. Interview with Jay Keasling in Durham, N.C. on April 16, 2011.

8. Jeanne Clark. et al. 2000. Patent Pools: A Solution to the Problem of Access in Biotechnology Patents? United States Patent and Trademark Office. 4-5.
9. David Resnick. A Biotechnology Patent Pool: An Idea Whose Time has Come? *Journal of Philosophy, Science & Law.* Vol. 3. 10. 2003.
10. Arti Rai and James Boyle. Synthetic Biology: Caught between Property Rights, the Public Domain, and the Commons. PloS Biology. 5(3):e58. 2007.
11. Clark. et al. 2000. 4-5.
12. Robert Carlson. 2010. Biology is Technology. Harvard University Press. 66-73.
13. Jeronimo Cello, Aniko Paul, and Eckard Wimmer. Chemical Synthesis of Poliovirus cDNA: Generation of Infectious Virus in the Absence of Natural Template. *Science.* 297(5583)1016-18. 2002.
14. Hamilton Smith et al. Generating a synthetic genome by whole genome assembly: phiX174 bacteriophage from synthetic oligonucleotides. PNAS. 100(26)15440-15445. 2003.
15. Terrence M. Tumpey et al. Characterization of the Reconstructed 1918 Spanish Influenza Pandemic Virus. *Science.* 310(5745)77-80. 2005.
16. Synthetic Genomics Vaccines Inc. Press Release. October 7, 2010. http://www.syntheticgenomics.com/media/press/100710.html
17. Drew Endy. Foundations for Engineering Biology. *Nature.* 438:449-453. 2005.
18. Endy. 2005. 451.
19. Michael B. Elowitz and Stanislas Leibler. A synthetic oscillatory network of transcriptional regulators. *Nature.* 403:336-338. 2000.
20. Ming-Tang Chen and Ron Weiss. Artificial cell-cell communication in yeast *Saccharomyces cerevisiae* using signaling elements from *Arabidopisis thaliana. Nature Biotechnology.* 23:1551-1555. 2005.
21. Maung Nyan Win and Christina Smolke. A modular and extensible RA-based gene regulatory platform for engineering cellular function. PNAS. 104(36)14283-14288. 2007.
22. Christina M. Agapakis et al. Insulation of a Synthetic Hydrogen Metabolism Circuit in Bacteria. *Journal of Biological Engineering.* 4:3. 2010. http://www.jbioleng.org/content/4/1/3
23. Carlson. 2010. 29-32.
24. Ed Pilkington. I am creating artificial life, declares US gene pioneer. *Guardian.* October 6, 2007.
25. Daniel G. Gibson et al. Creation of a Bacterial Cell Controlled by a Chemically Synthesized Genome. *Science.* 329(5987):52-56. 2010.
26. Coco Ballantyne. Longest Piece of Synthetic DNA Yet. *Scientific American.* 9. January 24, 2008.
27. Claire M. Fraser et al. The Minimal Gene Complement of Mycoplasma genitalium. Science. 270(5235):397-403. 1995.
28. Daniel G. Gibson et al. Complete Chemical Synthesis, Assembly, and Cloning of a *Mycoplasma genitalium* Genome. *Science.* 319(5867):1215-1220. 2008.
29. Gibson. 2010. 1.
30. J. Craig Venter Institute Press Release. J. Craig Venter Institute Researchers Clone and Engineer Bacterial Genomes in Yeast and Transplant Genomes Back into Bacterial Cells. August 20, 2009.

CHAPTER 11 THE BOO SCENARIOS

1. Extreme benefits is a parody in reference to ETC Group's 2007 article, *Extreme Genetic Engineering: An Introduction to Synthetic Biology.*
2. E-mail from Rob Carlson, May 6, 2010.
3. Michael Rogers. The Pandora's Box Congress. *Rolling Stone.* 82. June 1975.
4. James D. Watson. 2000. A Passion for DNA: Genes, Genomes, and Society. Cold Spring Harbor Laboratory Press. 68.
5. ETC Group. 2007. Extreme Genetic Engineering: An Introduction to Synthetic Biology. 47. http://www.etcgroup.org/en/node/602
6. Michael Yudell and Robert DeSalle. 2002. The Genomic Revolution: Unveiling the Unity of Life. Joseph Henry Press. 144.
7. Charles Q. Choi. *The Scientist.* Plant Pest Resistance Boosted. 2005. http://www.the-scientist.com/news/20050517/02
8. Steven Burrill. 2004. Biotech's Getting Back on Track: 18th Annual Report on the Industry. Burrill & Company. 278.
9. Tony Gilland. 2006. Trade War or Culture War? (Chapter 3 in Let Them Eat Precaution, Jon Entine editor). The AEI Press. 63-65.
10. Susan Dudley. Would You Eat Genetically Engineered Salmon? *The Daily Caller.* September 27, 2010.
11. Francois Jacob. Evolution and Tinkering. *Science.* 196(4295):1161-1166. 1977.
12. Andrew Ellington. On Artificial Proteins. *The Scientist.* 2011. http://www.the-scientist.com/news/display/57978
13. Jonathan B. Tucker and Craig Hooper. Protein Engineering: Security Implications. European Molecular Biology Organization. EMBO Reports. Vol. 7. 514-516. 2006.
14. Ronald J. Jackson et al. Expression of Mouse Interleukin-4 by a Recombinant Ectromelia Virus Suppresses Cytolytic Lymphocyte Responses and Overcomes Genetic Resistance to Mousepox. *Journal of Virology.* 75(3):1205–1210. 2001.
15. J. Cello, A.V. Paul, and E. Wimmer. Chemical Synthesis of Poliovirus cDNA: Generation of Infectious Virus in the Absence of Natural Template. *Science.* 297:1016-1018. 2002.
16. National Research Council. 2004. Seeking Security: Pathogens, Open Access, and Genome Databases. http://www.nap.edu/catalog.php?record_id=11087
17. Jiayuan Quan et al. Parallel on-chip gene synthesis and application to optimization of protein expression. *Nature Biotechnology.* April 2011.
18. Presidential Commission for the Study of Bioethical Issues. 2010. New Directions: The Ethics of Synthetic Biology and Emerging Technologies. 9. http://www.bioethics.gov
19. Michele Garfinkel, Drew Endy, Gerald Epstein, and Robert Friedman. 2006. *Synthetic Genomics: Options for Governance.* The Venter Institute, MIT, and the Center for Strategic and International Studies. 21.
20. E-mail correspondence between James Randerson and Drew Endy http://dspace.mit.edu/handle/1721.1/33001
21. James Randerson. Revealed: the lax laws that could allow assembly of deadly virus DNA. *The Guardian.* June 14, 2006. http://www.guardian.co.uk/world/2006/jun/14/terrorism.topstories3
22. Testimony by Jim Thomas representing ETC Group before the Presidential Commission for the Study of Bioethical Issues on July 8, 2010 in Washington, D.C.
23. ETC Group. Synthetic Biology- Global Societal Review Urgent. May 17, 2006 http://www.etcgroup.org/en/node/11

24. ETC Group News Release. NGOs Blast US Presidential Commission on Bioethics. December 16, 2010. http://etcgroup.org/en/node/5244

25. ETC Group News Release. Synthia is Alive and Breeding: Panacea or Pandora's Box? May 20, 2010. http://etcgroup.org/en/node/5142

26. White House Press Release. May 20, 2010. http://www.bioethics.gov/documents/Letter-from-President-Obama-05.20.10.pdf

27. Ethel Machi and Jena Baker McNeil. New Technologies, Future Weapons: Gene Sequencing and Synthetic Biology. The Heritage Foundation. 2. August 2010.

28. Testimony by Drew Endy representing Stanford University before the Presidential Commission for the Study of Bioethical Issues on July 8, 2010 in Washington, D.C.

29. J. Lazarou, BH Pomeranz, and PN Corey. Incidence of adverse drug reactions in hospitalized patients: a meta-analysis of prospective studies. *JAMA*. 279(15):1200-1205. 1998.

30. National Highway Traffic Safety Administration. List of Motor Vehicle Deaths in U.S. by Year. http://en.wikipedia.org/wiki/List_of_motor_vehicle_deaths_in_U.S._by_year

Index

1000 Genomes Project 20
Abzug, Bella 46
ACLU 46
Adler, Reid 45
Amgen 44, 45, 116
Applied Biosystems 4, 5, 7, 8, 10, 111, 141
Arber, Werner 87
Aristotle 56, 58, 79
Baker, Howard 27
Baltimore, David 4
Bayh-Dole Act 41, 44, 45, 47
Beadle, George 83
Becerra-Weldon 46
Behe, Michael 77, 166
Belmont Report 25
Berg, Paul 88
Bermuda Rules 6, 50
Bill and Melinda Gates Foundation 139
Biogen 45, 134
BLAST algorithm 17
blending inheritance 62
blood typing 19
Botstein, David 2
Boyden, Thomas 35
Boyer, Herbert 44, 45, 88, 89
Brenner, Sydney 17
Brown, Ron 28
Buck, Carrie 23
Buck v. Bell 23
Buffon, Comte de 57, 59, 68
Canavan disease 31
Carlson Curve 142
Carlson, Rob 10
Carlson, Robert iii, 10, 108, 141, 142, 144, 148, 162, 168, 170, 171, 172
Catalona, William 30
Cavalli-Sforza, Luca 28

Celera Genomics 7, 8, 9, 10, 92, 111, 112
Chain, Ernst 40
Chakrabarty, Ananda 48
Chambers, Robert 60, 62, 68
Chargaff, Erwin 84
Church, George 20
Clinton, Bill 9, 28, 34, 111, 120, 127, 148
Cohen, Stanley 44, 88
Collins, Francis 2, 6, 7, 9, 11, 15, 78, 89, 114, 169
Common Rule 28
Copernicus 56, 68
Council for Responsible Genetics 22
Crick, Francis 84
Cutler, John 25
cystic fibrosis gene 15
Darwin, Charles 56, 61, 62, 63, 66, 67, 68, 71, 76, 77, 79, 80, 96
Darwinism 66, 67, 68, 69, 76, 77, 129, 165
Dausset, Jean 19
Davenport, Charles 23
deCODE Genetics 33, 111, 112
DeLisi, Charles 3, 4
Dennett, Daniel 78
Department of Energy 2, 3, 4, 5, 9
Diamond v. Chakrabarty 49
Dobzhansky, Theodosius 78
Dulbecco, Renato 3
Duncan, John 27
Employee Retirement and Income Security Act 34
Endangered Species Act 26, 27
epigenetic regulation 73
Equal Protection Clause 24
Ethical, Legal, and Social Issues 10, 22, 35
Etnier, David 26

Eugenical Sterilization Act 23
eugenics 23
EvoDevo 73, 77, 131
ex-parte Allen 49
ex-parte Hibberd 47
expressed sequence tags 6, 10, 45, 46
Federal Transfer Technology Act 41
Fire, Andrew 95
Fleming, Alexander 40
Florey, Howard 40
Fourteenth Amendment 24
Franklin, Rosalind 84
G-5 labs 5
Galileo 13, 56, 57, 68
Galton, Francis 23, 62
GenBank 2, 9, 16, 50, 156
Genentech 44, 45, 116, 134
gene therapy 2
Genetic Bill of Rights 22
Genetic Information Nondiscrimination Act 34
Genographic Project 29
Geron 47, 48, 121, 122
Gilbert, Walter 4, 7, 8, 92, 134, 141
Greenberg v. Miami Children's Hospital Research Institute 31
Gulbrandsen, Carl 48
Haeckel, Ernst 72
Hagahai people 29
Hardy-Weinberg principle 69, 71
Harry, Debra 29
Hatch-Waxman Act 52
Havasupai Tribe v. Arizona Board of Regents 32
Health Sector Database Act 33
Healy, Bernadine 6
heterochrony 72
Hill, Hiram 26
HLA complex 19
Holmes, Oliver Wendell 24
Holt, Robert 9
Hood, Leroy 3, 141
Howard Hughes Medical Institute 5
Human Genome Diversity Project 28

Human Genome Organization 28
Human Genome Project 2, 3, 5, 9, 10, 11, 13, 15, 16, 20, 90, 107, 108, 109, 142, 168
Human Genome Sciences 7
Hunkapiller, Michael 5, 8
Huxley, Thomas 67, 69, 76
Incyte 46
Institute for Genome Research 7
interferon 44, 45, 134, 150
Jacob, Francois v, 93, 139, 154
Jacobson v. Massachusetts 24
Jeffreys, Alex 92
Joint Genome Institute 5, 9
Kent, James 9
Krugman, Paul 123, 132
Kuhn, Thomas 65, 165
Lamarck, Jean Baptiste Pierre v, 59, 62, 79, 99
Lander, Eric 9, 10, 12
Landsteiner, Karl 19
Laughlin, Harry 23
Leder, Philip 49
Lewontin, Richard 97, 130
Linnaeus, Carl 59, 68
Lipman, David 17
Madey v. Duke University 52
Martin, John 32
Matthew, Patrick 60
McClintock, Barbara 81, 87, 93
Medical Research Council 5
Mello, Craig 95
Mendel, Gregor 1, 21, 56, 62, 63, 67, 68, 80, 106
Merck v. Integra 52
microarray 14
Modern Synthesis 11, 69, 71, 72, 130
Monad, Jacques 93, 139
Moore v. Regents of the University of California 31
Morgan, Thomas Hunt 3, 17
Mouse Sequencing Consortium 16
Myers, Eugene 17
Myriad 46

Myriad Genetics 46

NASA 37

Nathans, Daniel 87

National Institutes of Health 2, 3, 4, 5, 6, 7, 16, 25, 29, 45, 46, 78, 107, 110, 111, 114, 115, 121, 133, 149

natural selection 62

Nazi Third Reich 24

Nixon, Richard 87

OncoMouse 49

pangenesis 62, 67

Parke-Davis v. H.K. Mulford 44

Pasteur, Louis 39, 40, 41, 42, 56, 58, 59, 68, 163

Patrinos, Ari 9

Pauling, Linus 85

penicillin 25, 39, 40, 41, 44, 107

Perkin Elmer 8

Personal Genome Project 20

Plant Patent Act 47

Plant Variety Protection Act 47

Plater, Zygmunt 26

restriction fragment length polymorphisms 15

Restriction fragment length polymorphisms 3

Reverby, Susan 25

Rifkin, Jeremy 46

Roche Pharmaceuticals 33, 51

Roche v. Bolar 51

Rockwell, Llewellyn H. 35

Sanger Centre 5

Sanger, Frederick 4, 8, 11, 141

Sarkar, Sahotra v, 73, 110

Scalia, Antonin v, 21

Schrodinger, Erwin 84

Schumpeter, Joseph 132, 170

Scopes Monkey Trial 76

Sexually Transmitted Disease Inoculation Study 25

shotgun sequencing 8, 10

Shreeve, James 7

single nucleotide polymorphisms 6, 15, 20, 21, 50, 104, 106, 107, 115

Sinsheimer, Robert 3

Skinner v. Oklahoma 24

Smith, Hamilton 8, 9, 87

SmithKline-Beecham 7

snail darter 25

spontaneous generation 56, 58

Stefansson, Kari 33, 111

Stevenson-Wydler Technology Innovation Act 41

Stewart, Timothy 49

Sulston, John 5, 6, 7, 9, 10, 45

Swanson, Robert 44

Tatum, Edward 83

Tellico Dam 25

Tennessee Valley Authority 25

Thirteenth Amendment 46

Thomson, James 47

Tsui, Lap-Chee 15, 16

Tuskegee syphilis study 25

TVA v. Hill 25

United States Patent and Trademark Office 6, 29, 39, 43, 45, 46, 47, 48, 49, 50, 111

Varmus, Harold 7

Venter, Craig 6, 7, 8, 9, 10, 19, 45, 89, 90, 97, 104, 111, 140, 142, 145, 146, 148, 157, 158, 171, 172

Wallace, Alfred 60, 61

Washington University v. Catalona 30

Waterston, Robert 6, 9

Watson, James 3, 5, 6, 9, 10, 83, 84, 107, 172

Weismann, August v, 59, 67, 81, 99

Wellcome Trust 5, 6

Wells, Spencer 29

Wells, William Charles 60

Whitehead, Irving 24

Wilkins, Maurice 84

Willard, Huntington 12

Winkler, Hans 1

Wisconsin Alumni Research Foundation 47, 48

Woese, Carl 96

Wyngaarden, James 3, 4

About the Author

Randall Mayes is a policy analyst specializing in biotechnology. His areas of expertise include technology based economic development and public policy issues related to genomics, nanotechnology, and synthetic biology including intellectual property, bioethics, risk assessment, and performance enhancement. From 2009-2010, he was a part of the Working Group that provided recommendations to the State of North Carolina regarding policies regulating and developing nanotechnology.

Mayes' first book, *The Cybernetics of Kenyan Running* (Carolina Academic Press 2005), investigates the biological, historical, cultural, and psychological components to Kenyan running success. He was the co-organizer and co-moderator of the Coming Age of the Uber-Athlete (2008) conference at the American Enterprise Institute in Washington. DC. He co-authored the media guide for Reebok International for the 2000 Sydney Olympics which profiles the coaches and track athletes in the Reebok Enclave.

Prior to analyzing biotechnology issues, Mayes was a free-lance journalist covering science as it relates to business, politics, and culture. He has also worked in business development for the technology and solar energy industries. The Rocky Mountain Institute Newsletter featured his work on solar powered web hosting.

Mayes has a Master of Environmental Management degree from the Nicholas School of the Environment and Earth Sciences at Duke University and guest lecturers on genomics and performance in sport. In 2009, he served as a fellow at the Institute for Ethics and Emerging Technologies. He is currently a senior fellow with STATS at George Mason University and resides in Durham, NC.

Related Titles from Logos Press®

http://www.logos-press.com

Building Biotechnology

Scientists know science; businesspeople know business. This book explains both.

Hardcover ISBN: 978-09734676-5-9
Softcover ISBN: 978-09734676-6-6

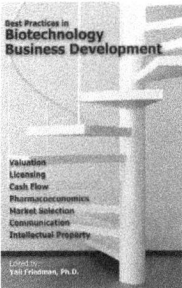

Best Practices in Biotechnology Business Development

Valuation, Licensing, Cash Flow, Pharmacoeconomics, Market Selection, Communication, and Intellectual Property

ISBN: 978-09734676-0-4

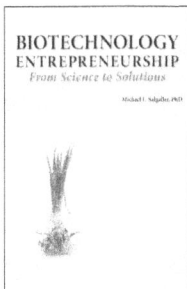

Biotechnology Entrepreneurship

From Science to Solutions

Hardcover ISBN: 978-1-934899-13-7
Softcover ISBN: 978-1-934899-14-4